HANDBOOK OF DRAFTING TECHNOLOGY

HANDBOOK OF DRAFTING TECHNOLOGY

JOHN A. NELSON

 **VAN NOSTRAND REINHOLD COMPANY**
NEW YORK   CINCINNATI   ATLANTA   DALLAS   SAN FRANCISCO
LONDON   TORONTO   MELBOURNE

Van Nostrand Reinhold Company Regional Offices:
New York   Cincinnati   Atlanta   Dallas   San Francisco

Van Nostrand Reinhold Company International Offices:
London   Toronto   Melbourne

Library of Congress Catalog Card Number: 80-20103
ISBN: 0-442-28661-9
ISBN: 0-442-28662-7 (pbk.)

Manufactured in the United States of America

Published by Van Nostrand Reinhold Company
135 West 50th Street, New York, N.Y. 10020

Published simultaneously in Canada by Van Nostrand Reinhold Ltd.

15 14 13 12 11 10 9 8 7 6 5 4 3 2 1

Library of Congress Cataloging in Publication Data

Nelson, John Augustus, 1935-
    Handbook of drafting technology.

    Includes index.
    1. Mechanical drawing.   I. Title.
T353.N46   604.2   80-20103
ISBN 0-442-28661-9
ISBN 0-442-28662-7 (pbk.)

PREFACE

Drafting is a communication skill that enables a designer, engineer, or architect to convey his or her ideas through the draftsperson to the skilled craftspersons who will ultimately do the building. From major calculations, general specifications and preliminary sketches of a project, the draftsperson must prepare an exact, detailed drawing. The job may include making computations from information in engineering handbooks and tables and, in some cases, actually describing the materials and processes to be used. Every facet of a project must be detailed with precision and clarity through the graphic language which is the essence of drafting.

Draftspersons are vital to virtually every phase of life in the United States. They make the drawings from which our homes, factories, and skyscrapers are built. Our roads and bridges, our refrigerators and washing machines start with their work. They make the drawings for our new automobiles—over 27,000 drawings for each new model!

The demand for draftspersons has been growing steadily. According to the U.S. Department of Labor, 250,000 draftspersons were employed in American industry on January 1, 1970. During 1975, the figure had risen to 290,000, and it is estimated that 385,000 will be needed in the 1980s. The term "draftsperson" includes women; nearly 12 percent of all draftspersons in the United States are women.

Those who intend to make drafting their career should include as much algebra, geometry, trigonometry and science as possible in their high school programs. The standards which must be met are high, but the rewards are great.

There are various fields of drafting: mechanical, electro-mechanical, architectural, electronic, civil and technical illustration are some of the major ones. The draftsperson usually specializes in one field in order to become truly competent. A skilled draftsperson in any field, however, must know and fully understand how the craftsperson will make whatever is drawn. In this way, and only in this way, can the draftsperson create a drawing that is both functional and feasible.

What are the chances for advancement in the field? For a good draftsperson, the opportunities are excellent. The draftsperson can move up the ladder from a junior to a senior position, to group leader, or to drafting supervisor. Related jobs in technical sales, purchasing, planning, marketing, and inspection are other possibilities for combining interests and skills. It is not unusual for a draftsperson to find success in this challenging and stimulating field which continues to expand, influencing so many aspects of our lives.

INTRODUCTION

This drafting handbook has been written to provide a nontechnical approach to the world of drafting. A special effort has been made to present all principles in illustrations and in concise descriptions which use words and phrases of the trade in a manner that will be understandable to all interested readers. Although technical principles are presented in varying degrees of complexity, the primary emphasis is placed on points of major importance.

This book has a three-fold purpose: to serve as a text for anyone wishing to gain an insight into the profession of drafting; to serve as a practical guide for the mechanical drafting student; and to function as a ready reference handbook for professional draftspersons, technical illustrators, technical writers, and engineers.

It is hoped that by reading and studying this text, the beginning student will also be able to read and understand the many kinds of blueprints used in industry today.

Personal thanks are extended to my wife, Joyce, for her moral support and untiring efforts in typing this material.

John A. Nelson

ACKNOWLEDGMENTS

The author wishes to thank the following for reviewing the manuscript and providing critical input:

Reviewers

Robert Franciose, Chairman
ANSI Drafting Standards (Y14) Committee

R. Michael Holcombe, Chairman
Drafting & Design Department
Asheville-Buncombe Technical Institute
Asheville, North Carolina

Delmar Staff

Industrial Education Editor — Mark W. Huth
Associate Editor — Kathleen E. Beiswenger
Technical Editor — Harry A. Sturges

Illustrations

Alvin Company
Berol USA
Engineering Graphics Title page photo
Keuffel & Esser Company
Teledyne Post
Vemco Corporation
L.S. Starrett Company

Classroom Testing

The instructional material in this text was classroom tested in the vocational drafting department at Conval Vocational High School — Peterborough, New Hampshire 03458.

CONTENTS

Preface	v
Introduction	vii
Unit 1 Equipment	1
Unit 2 Lettering	17
Unit 3 Drawing Techniques	21
Unit 4 Geometric Construction	25
Unit 5 Multiview Drawings	55
Unit 6 Basic Isometrics	71
Unit 7 Section Views	91
Unit 8 Descriptive Geometry	109
Unit 9 Auxiliary Views	121
Unit 10 Developments	125
Unit 11 Basic Dimensioning	149
Unit 12 Advanced Dimensioning	163
Unit 13 Manufacturing Processes	187
Unit 14 Basic Welding	203
Unit 15 Fasteners	213
Unit 16 Precision Measurement	237
Unit 17 Springs	247
Unit 18 Cams	255
Unit 19 Assembly and Detail Drawing	267
Unit 20 Mechanical Lettering	279
Unit 21 Perspective Drawing	291
Unit 22 The Engineering Department	307
Appendix A Inch/Metric Equivalents	333
Appendix B Circumferences and Areas	334
Appendix C Bend Allowances	335
Appendix D U.S. Standard Gauges of Sheet Metal	339
Appendix E Dimension And Size Chart For Threads	340
Appendix F Tables of Limits For Cylindrical Fits	341
Appendix G Drilled Hole Tolerance	345
Index	349

HANDBOOK OF DRAFTING TECHNOLOGY

UNIT 1

EQUIPMENT

SUGGESTED EQUIPMENT

Drafting equipment is very delicate and expensive. Extreme care must be used in adjusting, cleaning, using, and storing all instruments. Proper care of equipment is the responsibility of each student. In order to function in a professional manner, each student should have the following equipment.

Drafting table or drafting board
Drafting stool
Drafting machine or a T square
45° triangle
30°–60° triangle
Center wheel compass
Drop bow compass
Dividers
Mechanical, architectural, civil, and metric scales
Template assortment
French curves
Drafting brush
Dry cleaning pad
Protractor
Eraser
Erasing shield
Sandpaper paddle
Drafting tape
Pencils or lead holders

Drafting Machines

A drafting machine is a device which attaches to the drafting table and replaces the T square and triangles. The two types of drafting machines are the arm type and the track type, figure 1-1. On both types a round head holds two straightedges at right angles to one another. The head can be rotated to set the straightedges at any angle. Most machines are available with interchangeable straightedges marked with different scales along their edges.

A drafting machine is a precision instrument and should be checked for accuracy once a week. The instructions for checking and adjusting a drafting machine are included with the manufacturer's information.

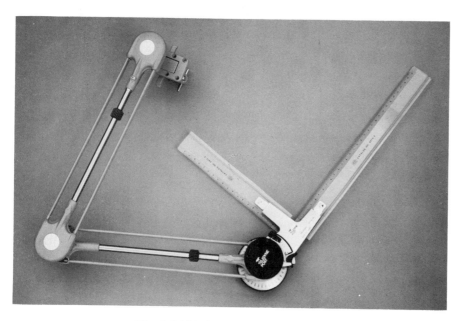

Fig. 1-1 (A) Arm-type drafting machine

Fig. 1-1 (B) Track-type drafting machine

T Square

The T square is used to draw horizontal lines, figure 1-2. Draw these lines only against the upper edge of the blade. Make sure the head is held securely against the left edge of the drawing board to guarantee parallel lines. This rule is followed as most drawing boards are not perfectly square. The left-handed drafter would reverse this procedure and place the head of the T square against the right edge of the board.

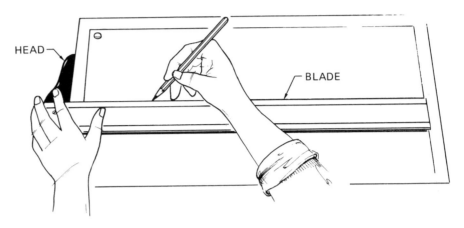

Fig. 1-2 Hold the T square head firmly against the drawing board edge with one hand. Draw horizontal lines with the other hand.

Parallel Straightedge

A parallel straightedge is sometimes used in place of a T square, figure 1-3. It is attached horizontally across the drawing board by a vertical wire threaded through both ends. This allows the straightedge to be moved up and down the drawing board and still remain parallel to the surface.

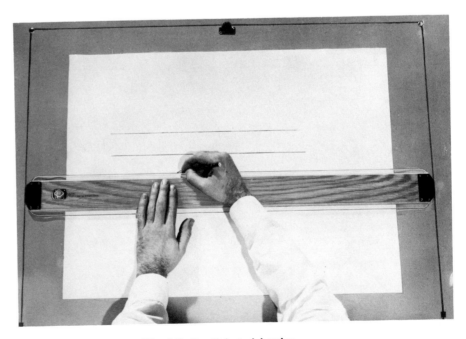

Fig. 1-3 Parallel straightedge

Triangles

There are two triangles used by drafters. One is called the 30–60-degree triangle, figure 1-4, usually written as 30°-60°. The other is a 45-degree triangle, figure 1-5, written as 45°. The 30°-60° contains a 30-degree, 60-degree, and 90-degree angle. The 45° consists of two 45-degree angles and one 90-degree angle.

Triangles are made of plastic and come in various sizes. When laying out lines, triangles are placed firmly against the upper edge of the T square. Pencils are placed against the left edge of the triangle and lines drawn upwards, away from the T square. Parallel angular lines are made by moving the triangle to the right after each new line has been drawn. Any angle divisible by 15 can be made by combining the 30°-60° and 45° triangles, figure 1-6.

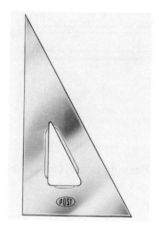

Fig. 1-4 30°-60° triangle

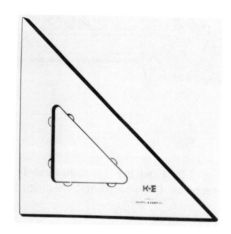

Fig. 1-5 45° triangle

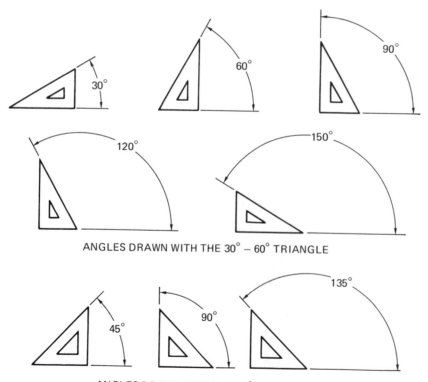

ANGLES DRAWN WITH THE 30° – 60° TRIANGLE

ANGLES DRAWN WITH THE 45° TRIANGLE

Fig. 1-6 Triangles

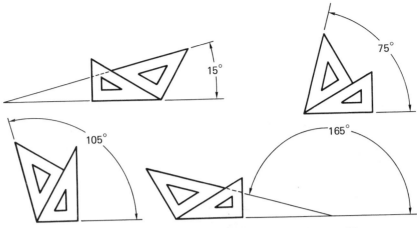

ANGLES DRAWN WITH THE COMBINED TRIANGLES

Adjustable Triangle

An adjustable triangle may take the place of both the 30°–60° and 45° triangles, figure 1-7. It is recommended, however, that this tool be used for drawing angles that cannot be made with the two standard triangles. The adjustable triangle is set by eye and is, therefore, not as accurate as the solid triangle.

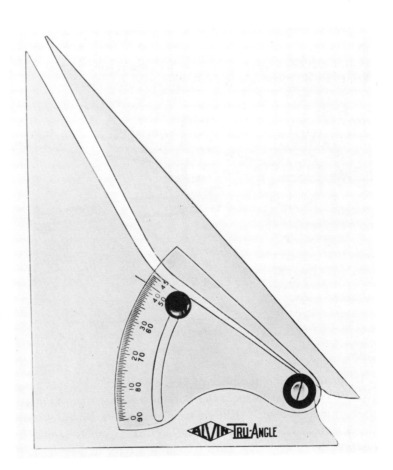

Fig. 1-7 Adjustable triangle

Drawing Instrument Sets

Typical drawing sets include compasses, dividers, and ruling pen, figure 1-8. Many sets include a variety of tools not normally used by a drafting student. It is recommended that only those tools actually needed be purchased.

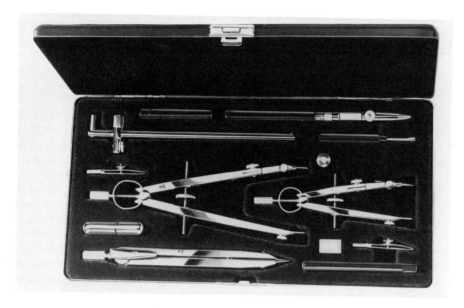

Fig. 1-8 Drawing instrument set

Divider

A divider is like a compass except it has a metal point on each leg, figure 1-9. It is used to lay off distances and to transfer measurements.

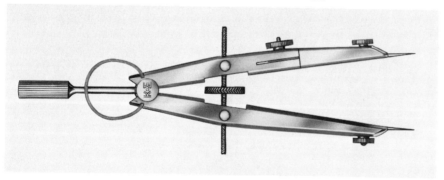

Fig. 1-9 Divider

Templates

A template is a thin, flat piece of plastic with various shapes cut in it, figure 1-10. It is designed to speed the work of the drafter and to make the finished drawing more accurate. There are templates to draw circles, ellipses, plumbing fixtures, nuts and bolts, screw threads, electronic symbols, springs, gears, and structural metals, to name just a few. Templates come in many sizes to fit the scale being used on the drawing.

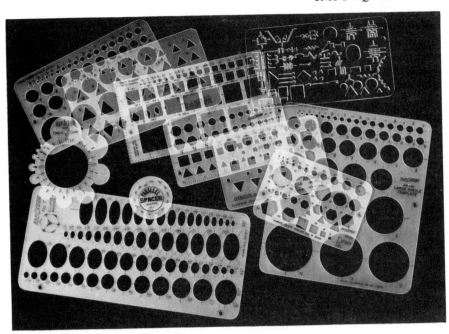

Fig. 1-10 Templates

French Curves

French curves are thin, plastic tools which come in an assortment of curved surfaces, figure 1-11. They are used to produce curved lines that cannot be made with a compass. Such lines are referred to as *irregular curves.*

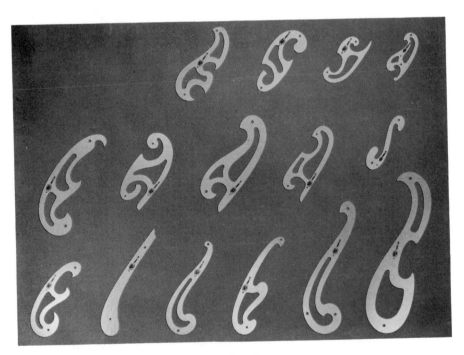

Fig. 1-11 French curves

Protractor

A protractor is used to measure and lay out angles, figure 1-12.

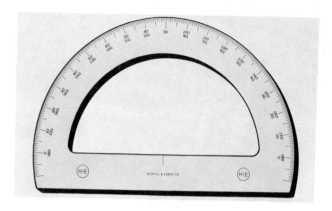

Fig. 1-12 Protractor

Pencils

Pencils come in 18 degrees of hardness ranging from 9H, which is very hard, to 7B, which is very soft, figure 1-13. The scale of hardness is as follows:

9H 8H 7H 6H 5H 4H	3H 2H H F HB B	2B 3B 4B 5B 6B 7B
Hard	Medium	Soft
accuracy	general purpose	art work

4H lead is recommended for layout work, extension lines, dimension lines, center lines, and section lines. 2H lead is used for object or visible edge lines and hidden lines. One should experiment with various leads to determine which lead gives the best line thickness. This varies depending on the pressure applied to the point while drawing lines.

Pencils are sharpened with a pencil sharpener. The important thing is that enough wood is removed to ensure that the lead, not the wood, of the pencil comes in contact with the T square or triangle edge.

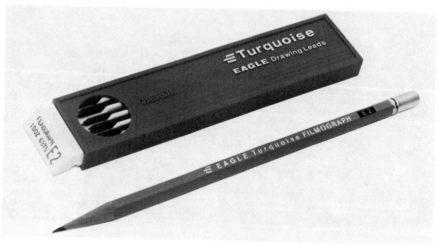

Fig. 1-13 Drawing pencil

Lead Holders and Leads

Lead holders hold sticks of lead, figure 1-14. The leads designed for lead holders come in the same range of hardness and are used for the same purposes as regular mechanical pencils. The main advantage is that they are more convenient to use. Leads are usually sharpened in a lead pointer, figure 1-15, or on a sandpaper paddle.

Fig. 1-14 Lead holders

Fig. 1-15 Lead pointer

Erasing Shield

An erasing shield restricts the erasing area, figure 1-16. In this way correctly drawn lines will not be disturbed during the erasing procedure. It is made from a thin, flat piece of metal with various size holes cut in it. To use, place the shield over the line to be erased and erase through the shield.

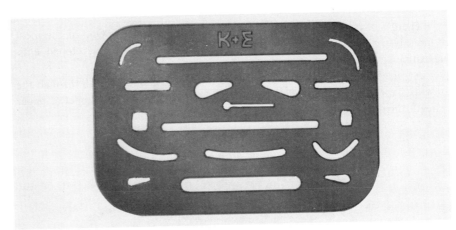

Fig. 1-16 Erasing shield

Drafting Brush

The drafting brush is used to remove loose graphite and eraser crumbs from the drawing surface, figure 1-17. Do not brush off a drawing surface by hand as it tends to smudge the drawing.

Fig. 1-17 Drafting brushes

Dry Cleaning Pad

A dry cleaning pad is used to erase minor smudges from the drawing surface, figure 1-18. Extreme care should be used as improper use of the dry cleaning pad will dull the lines of the drawing.

Fig. 1-18 Dry cleaning pad

Erasers

There are various kinds of erasers available to a drafter. One of the most commonly used is a soft, white eraser. If good drawing habits are developed, erasing can be kept to a minimum.

An electric eraser saves time, but care must be taken not to rub through the drawing paper, figure 1-19. This can be avoided by placing a thick sheet of paper beneath the drawing to cushion it.

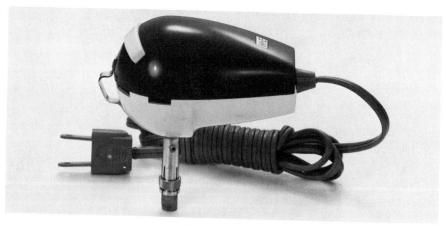

Fig. 1-19 Electric eraser

Proportional Dividers

Proportional dividers are used to enlarge or reduce an object in scale, figure 1-20. This tool has a sliding, adjustable pivot which varies the proportions of the tips of each leg.

Fig. 1-20 Proportional dividers

Sandpaper Paddle

A sandpaper paddle consists of several layers of sandpaper attached to a small wooden holder, figure 1-21. The sandpaper is used to sharpen pencil and lead points. Do not sharpen leads over a drawing as the graphite will smear the drawing surface.

Fig. 1-21 Sandpaper paddle

Compasses

There are two main types of compasses, figure 1-22. One is the friction-joint type and the other is the spring-bow type. The *friction-joint type* is still widely used for lightly laying out pencil drawings which will be inked. The disadvantage of this type compass is that the setting may slip when strong pressure is applied to the lead.

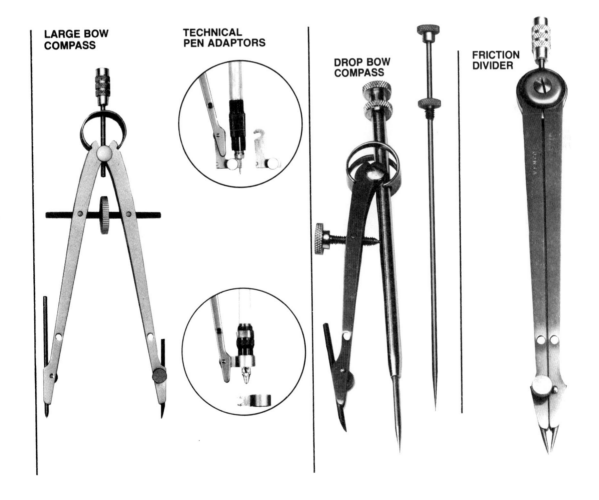

Fig. 1-22 Compasses

The *spring-bow type* is best for pencil drawings and tracings as it retains its setting even when strong pressure is applied to obtain dark lines. The spring, located at the top of the compass, holds the legs securely against the adjusting screw. The adjusting screw is used to make fine adjustments.

Compass leads should extend approximately 3/8 inch (9). The metal point of the compass is extended slightly more than the lead to compensate for the distance it enters the paper. The lead is sharpened with a sandpaper paddle to produce clean, sharp lines. The flat side of the lead faces outward in order to produce very small diameter circles, figures 1-23 and 1-24.

The compass is revolved between the thumb and the index finger. Pressure is applied downward on the metal point to prevent the compass from jumping out of the center hole, figure 1-25.

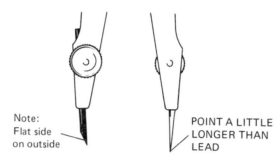

Note:
Flat side
on outside

POINT A LITTLE
LONGER THAN
LEAD

Fig. 1-23

SHARPEN WITH
PADDEL IN
DIRECTION OF
ARROW

Fig. 1-24

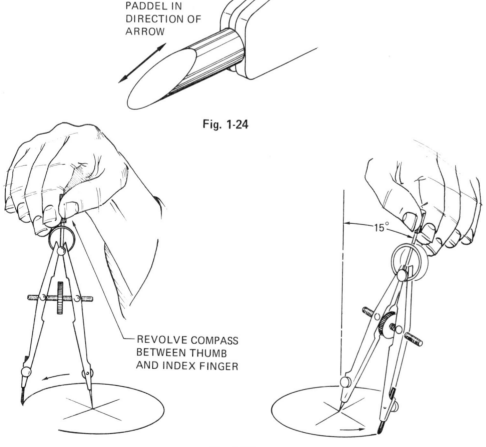

REVOLVE COMPASS
BETWEEN THUMB
AND INDEX FINGER

15°

Fig. 1-25

Paper Sizes

Paper sizes come in two standard increments in inches. All sizes fold up to the basic A size paper of 8 1/2 x 11 inches or 9 x 12 inches. Paper also comes in standard metric sizes and folds up to the basic A-4 size of 210 x 297 millimetres, figures 1-26 and 1-27.

INCHES		MILLIMETRES	
SIZE	DIMENSIONS	SIZE	DIMENSIONS
A	8 1/2 x 11 9 x 12	A-4	210 x 297
B	11 x 17 12 x 18	A-3	297 x 420
C	17 x 22 18 x 24	A-2	420 x 594
D	22 x 34 24 x 36	A-1	594 x 841
E	34 x 44 36 x 48	A-0	841 x 1189

Fig. 1-26

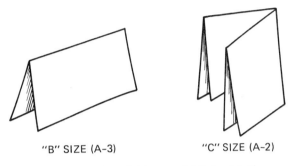

"B" SIZE (A-3)   "C" SIZE (A-2)

Fig. 1-27 All sizes fold up to "A" size (A-4)

Scales

There are various kinds of scales used by drafters, figure 1-28. A number of different scales are included on each instrument. They save the drafter the work of computing new measurements every time a drawing is made larger or smaller than the original.

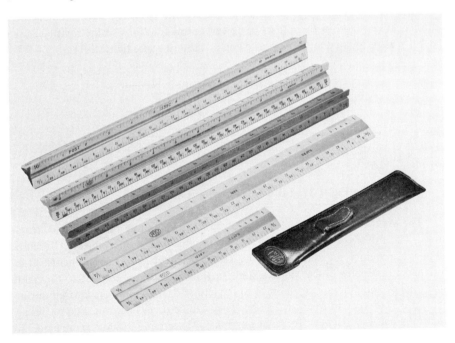

Fig. 1-28 Scales

Scales come open divided and full divided. A *full-divided scale* is one in which the units of measurement are subdivided throughout the length of the scale. An *open-divided scale* has its first unit of measurement subdivided, but the remaining units are open or free from subdivision.

Mechanical engineer's scales are divided into inches and parts of an inch. To lay out a full-size measurement, use the scale marked 16. This scale has each inch divided into 16 equal parts, or divisions of 1/16 inch. To use, place 0 on the point where measurement begins and step off the desired length, figure 1-29.

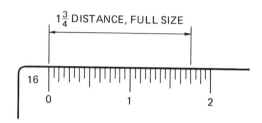

Fig. 1-29 1 3/4" measured out at full scale

To reduce a drawing 50 percent, use the scale marked 1/2. The large 0 at the end of the first subdivided measurement lines up with the other unit measurements that are part of the same scale. The large numbers crossed out in figure 1-30 go with the 1/4 scale starting at the other end. These numbers are ignored while using the 1/2 scale. To lay out 1 3/4 inches at the 1/2 scale, read full inches to the right of 0 and fractions to the left of 0.

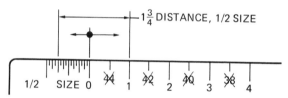

Fig. 1-30 1 3/4" measured out at 1/2 scale

The 1/4 scale is used in the same manner as the 1/2 scale. Measurements of full inches are made to the left of 0, however, and fractions to the right because the 1/4 scale is located at the opposite end of the 1/2 scale, figure 1-31.

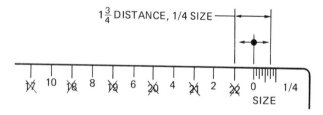

Fig. 1-31 1 3/4" measured out at 1/4 scale

The *architect's scale* is used primarily for drawing large buildings and structures. The full-size scale is used frequently for drawing smaller objects. Because of this, the architect's scale is generally used for all types of measurements. It is designed to measure in feet, inches, and fractions. Measure full feet to the right of 0, inches and fractions of an inch to the left of 0. The numbers crossed out in figure 1-32 correspond to the 1/2 scale. They can be used, however, as 6 inches as each falls halfway between full-foot divisions. Measurements from 0 are made in the opposite direction of the full scale because the 1/2 scale is located at the opposite end of the scale, figure 1-33.

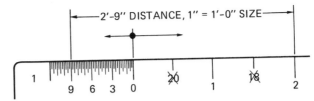

Fig. 1-32 2'-9" measured out at full scale

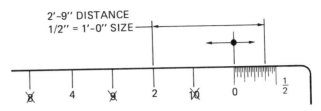

Fig. 1-33 2'-9" measured out at 1/2 scale

A *civil engineer's scale* is also called a *decimal inch scale.* The number 10, located in the corner of the scale in figure 1-34, indicates that each graduation is equal to 1/10 of an inch or .1". Measurements are read directly from the scale. The number 20, located in the corner of the scale shown in figure 1-35, indicates that it is 1/2 scale. To read 1/4 scale, the #40 scale (not shown) would be used.

A metric scale is used if the millimetre is the unit of linear measurement. It is read the same as the decimal scale except it is in millimetres, figure 1-38.

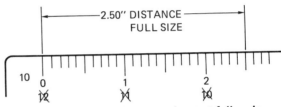

Fig. 1-34 2.50" measured out at full scale

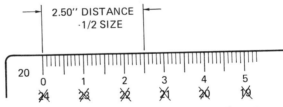

Fig. 1-35 2.50" measured out at 1/2 scale

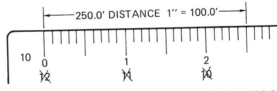

Fig. 1-36 250.0' measured out on scale 1" – 100.0'

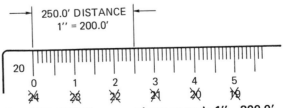

Fig. 1-37 250.0' measured out on scale 1" – 200.0'

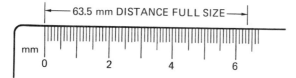

Fig. 1-38 63.5 mm measured out at full scale

WHITEPRINTERS

There are many types of whiteprinters available for use in drafting rooms. These machines reproduce a drawing through a chemical process. Most work on basically the same principle. A bright light passes through the translucent original drawing and onto a coated whiteprint paper. The light breaks down the coating on the whiteprint paper, but wherever lines have been drawn on the original drawing, no light strikes the coated sheet. Then the whiteprint paper is passed through ammonia vapor for developing. This chemical developing causes the unexposed areas — those which were shaded by lines on the original — to turn blue or black, figure 1-39.

On most whiteprinters there are controls to regulate the speed and the flow of the developing chemical. Each type of machine requires different settings and has different controls. Before operating any whiteprinter, read all of the manufacturer's instructions.

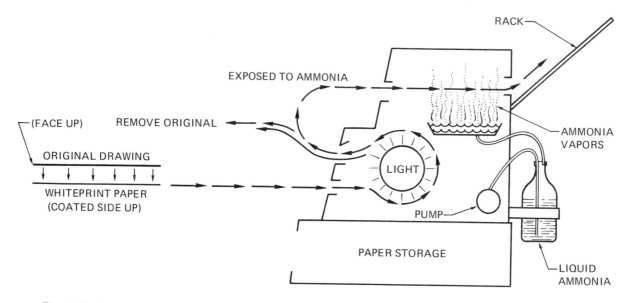

Fig. 1-39 Basic operating principles of ammonia-vapor whiteprinter

UNIT 2

LETTERING

LETTERING

Lettering is a skill that every drafter must perfect. Lettering on a drawing must be easy to read and understand. With practice and a knowledge of the proper order of strokes, anyone can learn to make uniform, legible letters.

The single stroke, uppercase Gothic letter is the most frequently used style of lettering on drawings. These letters are drawn either on a slant or vertically straight. Only one style is used on a drawing, however.

Whichever style is chosen, all letters should appear to have the same height, proportion, inclination, weight, and spacing between letters and words. Drawing guidelines is one way to help control lettering.

Guidelines

Guidelines help keep the height and inclination of letters uniform. They are drawn very lightly with a hard (4H or 6H) pencil, so lightly that they need not be erased. The light horizontal lines are drawn 1/8 inch (3) apart. Vertical or inclined lines are added to help space the letters, figure 2-1.

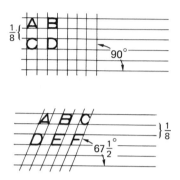

Fig. 2-1 Horizontal guidelines are spaced 1/8 inch (3) apart. Vertical and inclined guidelines help in spacing and proportioning letters.

Order of Strokes

Knowing the proper order of strokes makes it easier to draw each letter. When lettering, there are no upward strokes. Every stroke is downward, with the exception of horizontal lines which are drawn to the right, figure 2-2.

The order of strokes and proportion of inclined letters are the same as those for vertical letters. With inclined letters, the curved lines are more elliptical than circular, and straight lines lean to the right.

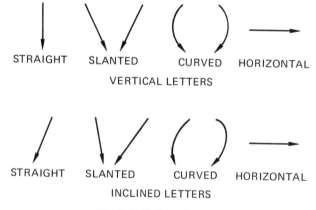

Fig. 2-2 Order of strokes

Study how letters are drawn in figure 2-3, both vertical and inclined styles. Notice how each is well-balanced and in proportion to the other letters. Numbers are drawn in the same proportion as letters. Fractions are drawn 3/8 inch (9) high. Use a sharp 2H lead pencil for lettering.

Left-handers may find that the order of strokes recommended do not work for them. These students should consult with their instructor and experiment with the order of strokes until each letter can be drawn smoothly, legibly, and in proper proportion.

Vertical style

Inclined style

Fig. 2-3 How to form uppercase Gothic letters and numerals

Spacing of Letters

Although the space between letters of a word is not actually equal, it must appear to be, figure 2-4. Improperly spaced letters make the text very difficult to read. It requires a great deal of practice to learn proper spacing, but the finished work will make the effort worthwhile.

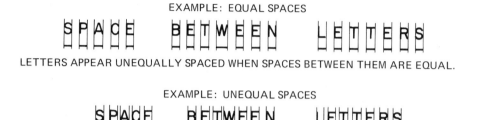

EXAMPLE: EQUAL SPACES

SPACE BETWEEN LETTERS

LETTERS APPEAR UNEQUALLY SPACED WHEN SPACES BETWEEN THEM ARE EQUAL.

EXAMPLE: UNEQUAL SPACES

SPACE BETWEEN LETTERS

LETTERS ARE SPACED SO THEY APPEAR TO HAVE THE SAME AMOUNT OF SPACE BETWEEN THEM.

Fig. 2-4 Spacing of letters

Word and Sentence Spacing

- The area between *words* = the height of the letters
- The area between *sentences* = twice the height of the letters
- The area between *lines* = from 1/2 to 1 1/2 times the height of the letters

BORDERS

A drawing must have a frame around it to give it a good appearance. This frame is referred to as a *border*, figure 2-5. Borders should have the following measurements:

PAPER SIZE	LEFT	TOP, BOTTOM, AND RIGHT
A & B	1 1/4"	1/4"
C	1 1/4"	3/8"
D	1 1/4"	1/2"
A4 & A3	30 mm	6 mm
A2	30 mm	9 mm
A1	30 mm	12 mm

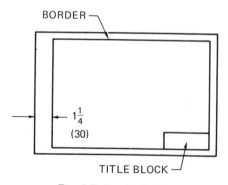

BORDER

$1\frac{1}{4}$
(30)

TITLE BLOCK

Fig. 2-5 Border layout

UNIT 3

DRAWING TECHNIQUES

WORK HABITS

To become an efficient drafter, the beginning student should have average or above average talent in mechanical reasoning, finger dexterity, and artistic ability. The student's competency in each area must be continually improved by repetition of exercises. In order to gain speed, neatness, and accuracy, it is important to develop good work habits. Many habits, such as keeping equipment neatly stored, clean, and ready for immediate use, are very basic work habits. Other, more involved habits to develop include using the correct steps in centering, laying out, and darkening drawings.

Correct work habits and procedures are emphasized throughout this text. Practice each so that drawings will be made quickly, neatly, and accurately. Do not take shortcuts or the finished drawing will suffer as a result.

ALPHABET OF LINES

Each line used to make a mechanical drawing has its own meaning. These lines of various shapes and weights (thicknesses) are called an *alphabet of lines.* As drawings become more complicated, it is most important that lines follow the suggested shape and weight illustrated in figure 3-1. Alphabet of lines should all be the same blackness. They differ only in line thickness.

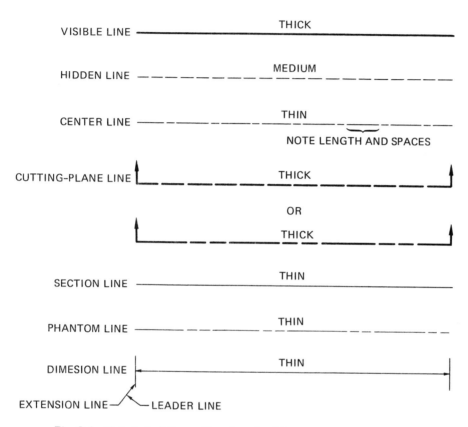

Fig. 3-1 Alphabet of lines. Note length of lines, dashes, and spaces.

SUGGESTED DRAWING STEPS

Steps one through seven show how to draw an object. Note which pencil leads are used to draw the various lines. Figure 3-2 shows the object to be drawn.

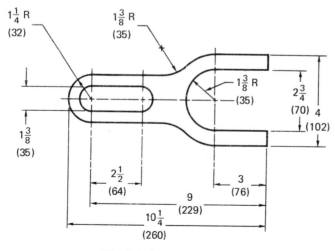

Fig. 3-2 Object to be drawn

Step 1. Locate the horizontal center line and center the basic shape using light lines made with a 4H lead.

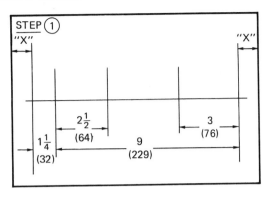

Step 2. Lightly locate all arcs and radii using a 4H lead.

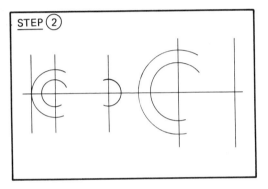

Step 3. Construct all straight lines lightly with a 4H lead. Check dimensions.

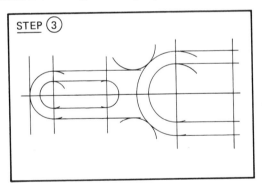

Step 4. Thicken all arcs and radii using a 2H lead.

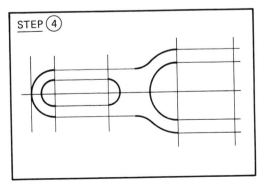

Step 5. Draw thin, black center lines using a 4H lead.

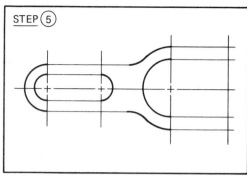

Step 6. Thicken all straight lines with a 2H lead.

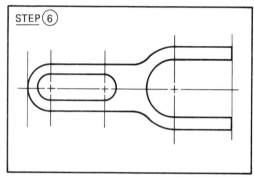

Step 7. Check dimensions, centering, line work and overall neatness.

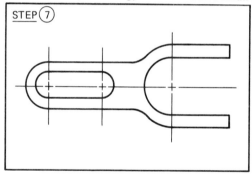

THINGS TO AVOID

Much will be said about developing good drafting habits. The following are considered *poor* drafting practice:

- Do not use a scale as a straightedge for drawing lines.
- Do not draw horizontal lines with the lower edge of the straightedge.
- Do not cut paper with a knife using the edge of a straightedge or triangle as a guide.
- Do not work with an unsharpened pencil.
- Do not sharpen a compass lead over the drawing board.
- Do not jab the dividers or compass points into the drawing board.
- Do not draw a circle or radius with a compass unless the point is sharp and extends 3/8 inch (9) from the edge of the compass.
- Do not use dividers as pincers or picks.
- Do not start any drawing unless all equipment is clean.
- Do not fill a drawing pen over a drawing.
- Do not put a compass away without opening it to relieve the spring tension.

UNIT 4

GEOMETRIC CONSTRUCTION

GEOMETRIC CONSTRUCTION

Knowing how to construct geometric shapes helps a drafter to make accurate drawings. This unit explains the terms and principles of geometry through simple geometric constructions.

When laying out a geometric construction, use very light lines made with a 4H pencil lead. When the construction is completed, use a 2H lead to darken the shape with thin, black lines.

Figure 4-1 illustrates some common geometric shapes.

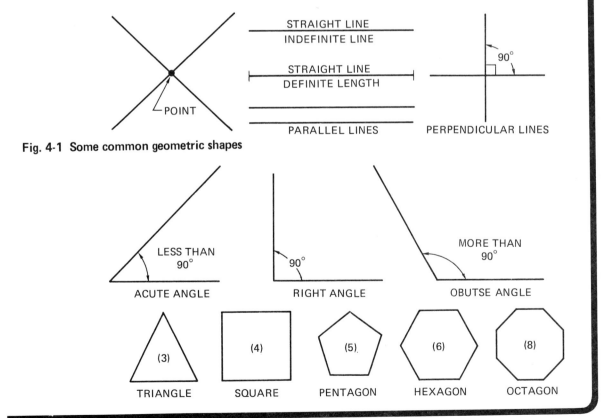

Fig. 4-1 Some common geometric shapes

HOW TO BISECT A LINE

- *Bisect* means to cut in half.
- Where two lines cross is their point of *intersection*.
- *Perpendicular* means at right angle (90°).

Given:

LINE AB

Step 1. Set the compass point at approximately two-thirds the length of line AB and swing an arc from point A.

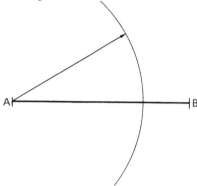

Step 2. Using the same compass setting, swing an arc from point B.

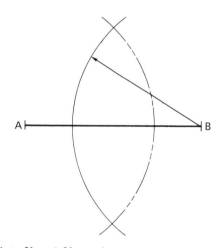

Step 3. Points X and Y are formed by the intersection of arcs A and B. Draw a line connecting points X and Y. The line connecting X and Y is perpendicular to and bisects given line AB.

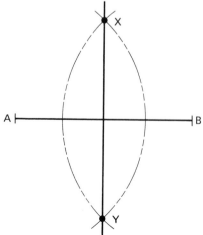

HOW TO BISECT AN ANGLE

Given:

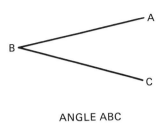

ANGLE ABC

Step 1. Set the compass point at any convenient radius and swing an arc from point B.

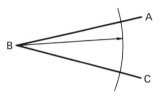

Step 2. Points X and Y are formed where the arc crosses lines A and C. Swing two identical arcs from points X and Y.

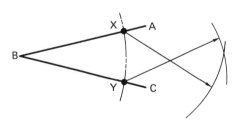

Step 3. Point Z is formed where the arcs from X and Y cross. Draw a line from B to Z. Line BZ bisects angle ABC.

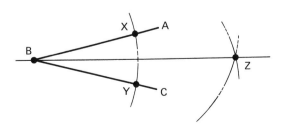

HOW TO SWING AN ARC OR CIRCLE THROUGH THREE GIVEN POINTS

Given:

B
+

+ C

A +

POINTS A, B, C

Step 1. Connect points B and C with a straight line. Connect points A and B with a straight line.

Step 2. Using the method outlined for bisecting a line, bisect lines AB and BC.

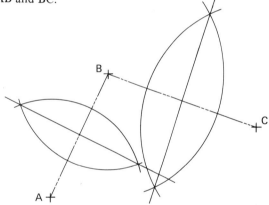

Step 3. Point X is formed where the two extended bisectors meet. Point X is the center of arc or circle ABC which is drawn using radius XA.

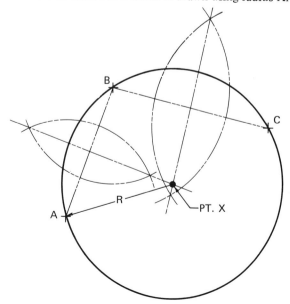

HOW TO LOCATE THE CENTER OF A CIRCLE

Given:

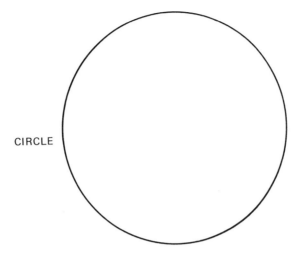

CIRCLE

Step 1. Using a drafting machine or T square, draw a horizontal line across the circle at a place approximately halfway from the top to the center of the circle. Where this line passes through the circle forms points A and B.

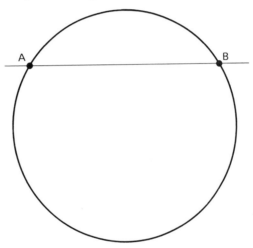

Step 2. Draw perpendicular lines downward from points A and B. Where these lines cross the circle forms points C and D.

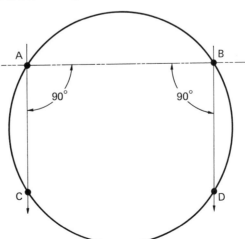

Step 3. Carefully draw a line from C to B and from A to D. Where these lines cross is the exact center of the given circle.

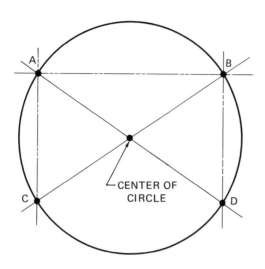

HOW TO DRAW A HEXAGON

Given:

Distance across the flats, or surface, of the desired hexagon

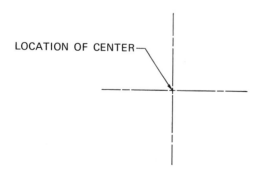

LOCATION OF CENTER

Step 1. Draw a circle with a diameter equal to the distance across the flats of the desired hexagon.

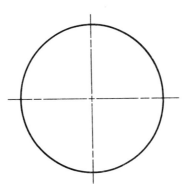

Step 2. Draw two horizontal lines (or two vertical lines if the hexagon is to be illustrated in another position) tangent to the circle.

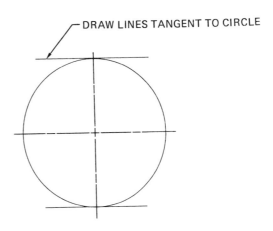

Step 3. Using a 30°-60° triangle, complete the hexagon as shown. Use the correct line weight. Darken the hexagon.

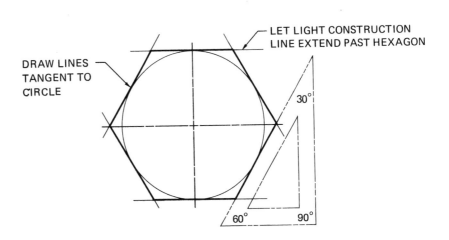

HOW TO DIVIDE A LINE INTO EQUAL PARTS

Given:

LINE AB

Problem: To divide line AB into three equal parts:

Step 1. Draw a line 90 degrees from either end of the given line.

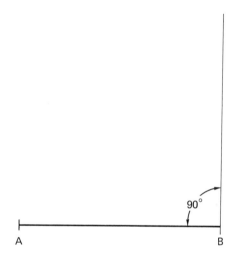

Step 2. Place the scale with its 0 on point A of the given line. Pivot the scale until the 3-inch measurement, or any multiple of 3 units of measure, is on the perpendicular line drawn in Step 1.

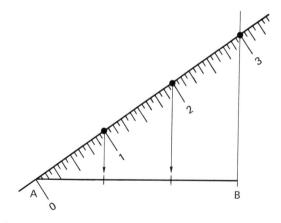

Step 3. Place dots at points 1 and 2 as shown. Project lines 90 degrees downward from these points. Add hash marks where projected lines cross given line AB.

HOW TO TRANSFER ODD SHAPES

Method I

> *Given:* TRIANGLE ABC

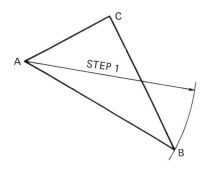

Step 1. Place the compass point at letter A, extend lead to letter B. Draw a light arc at the desired location. Letter the point A. Fix letter B at any convenient place on arc. (Letter each point as you proceed.)

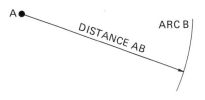

Step 2. Place the compass point at letter B of the original drawing and extend compass lead to letter C. Transfer distance BC as illustrated.

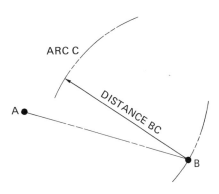

Step 3. Place compass point at letter A of the original drawing and extend compass lead to letter C. Transfer distance AC as illustrated.

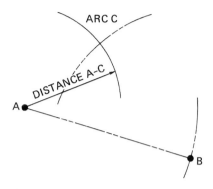

Step 4. Locate and letter each point.

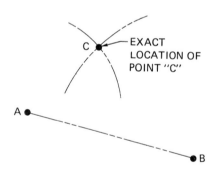

Step 5. Connect points A, B, and C. Original shape has been transferred.

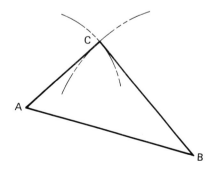

HOW TO TRANSFER ODD SHAPES

Method II

Given:

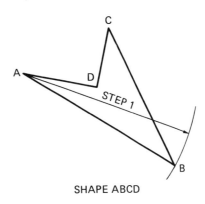

SHAPE ABCD

Step 1. Using the same procedure as used to transfer a three-sided, odd-shaped figure, locate and transfer distance AB.

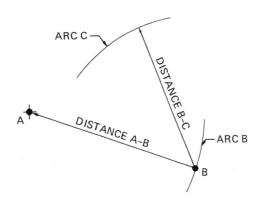

Step 2. Fix point B at any convenient place on arc B. Measure and transfer distance BC.

Step 3. Measure and transfer distance AC. Locate point C where distances AB and BC cross. Letter point C.

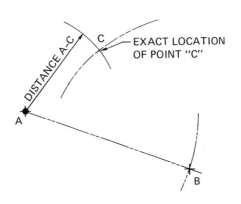

Step 4. Measure and transfer distances AD and CD. Point D is where arcs AD and CD cross. Letter point D.

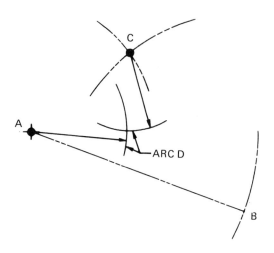

Step 5. Connect points AB, BC, CD, and DA. The shape has been transferred.

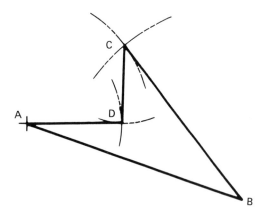

HOW TO TRANSFER COMPLEX ODD SHAPES

Given:

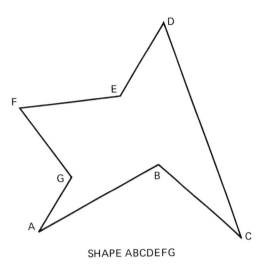

SHAPE ABCDEFG

The transferring of more complex shapes is easier if the basic shape is seen as a series of triangles.

The first step is to lay out the longest triangle contained within the basic shape, in this case ADC. Once the longest triangle is transferred, continue transferring all other points using the same procedure as in previous exercises. It will ease the transfer problem even more if letters or numbers are placed around the object in a systematic order.

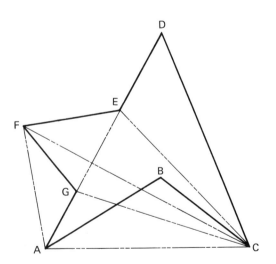

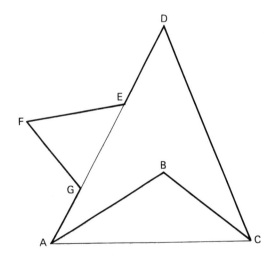

HOW TO LOCATE TANGENT POINTS

A *tangent point* is the exact point where one line stops and another line starts. *Tangent* means to touch. As an example, a tangent point is the exact point where a straight line stops being a straight line, and a curved line starts.

Tangent points are projected from the compass swing point at 90 degrees to the straight line next to it. Place a hash mark at each tangent point on the basic layout of each drawing, A, B, C, and D. This light layout work is done with a sharp 4H pencil lead.

Drawing A

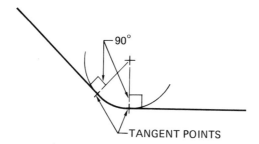

Drawing B

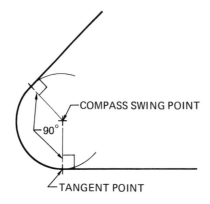

Drawing C

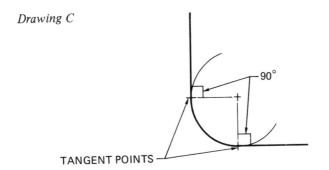

TANGENT POINTS

Drawing D

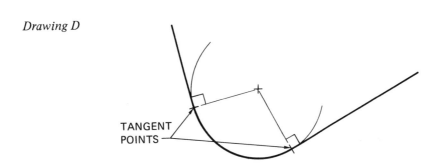

TANGENT
POINTS

When one arc blends into another arc, the tangent point is found by drawing a light line from one swing point to the next swing point. The tangent point is where this line crosses the arc. Add a hash mark at each point.

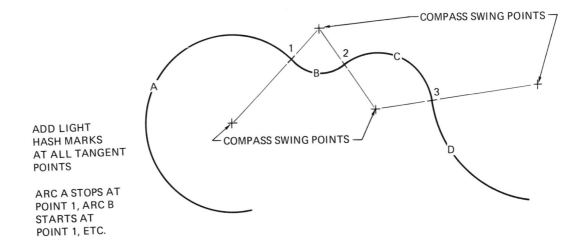

COMPASS SWING POINTS

COMPASS SWING POINTS

ADD LIGHT
HASH MARKS
AT ALL TANGENT
POINTS

ARC A STOPS AT
POINT 1, ARC B
STARTS AT
POINT 1, ETC.

HOW TO DRAW AN ELLIPSE

Given:

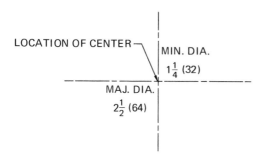

Step 1. Lightly draw a circle the size of the major diameter given. Using the same center, lightly draw a circle the size of the minor diameter. Number points 3, 6, 9, and 12.

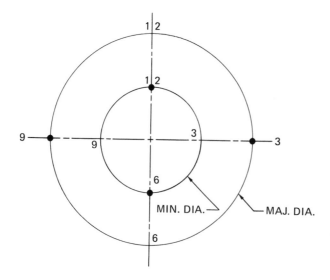

Step 2. Divide the given circles into 12 equal parts. Use a 30°–60° triangle as 12 x 30° = 360°, which is the total number of degrees in a circle. Number each point in order, clockwise around the circle.

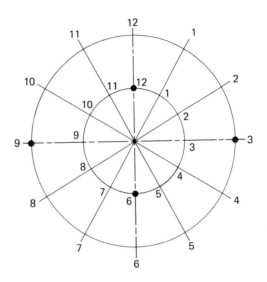

Step 3. Project at 90 degrees downwards from points, 10, 11, 1, and 2 located on the major diameter above the center line. Project at 90 degrees upwards from points 8, 7, 5, and 4 located on the major diameter below the center-line.

Project horizontal lines from points located on the minor diameter. These lines are projected until they meet the corresponding numbered lines projected from major diameter. The points of the ellipse are formed by the intersections.

Form the ellipse by connecting all newly formed points with french curves.

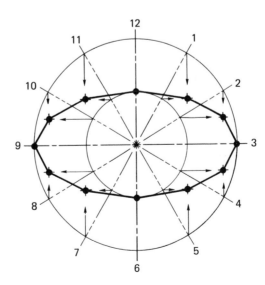

HOW TO DRAW AN OGEE CURVE

Method I. Both Parts of Curve Have Same Radius

Given:

LINES AB AND CD

Step 1. Draw a line from B to C. Bisect line BC to find point X.

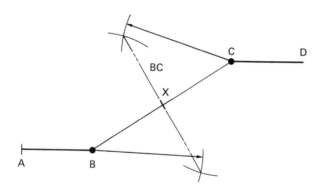

Step 2. Bisect line BX.

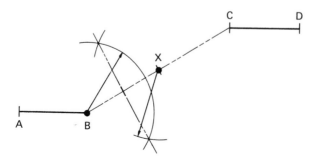

Step 3. Bisect line XC.

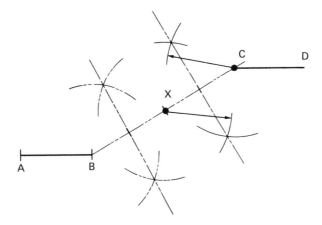

Step 4. Project 90 degrees upwards from point B and 90 degrees downwards from point C to bisect line and locate swing points.

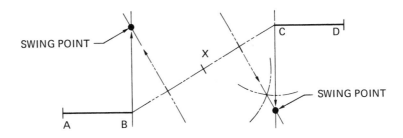

Step 5. Swing arcs BX and XC from swing points. Use hash marks to indicate tangent points.

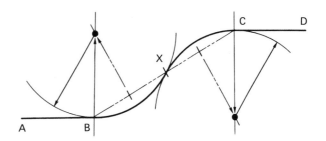

HOW TO DRAW AN OGEE CURVE

Method II. Parts of Curve Unequal

Given:

LINES AB AND CD

Problem: To place point X two-thirds the distance *from* B on line BC.

Step 1. Draw line BC. Divide line BC into three equal parts. Place point X on second division from point B.

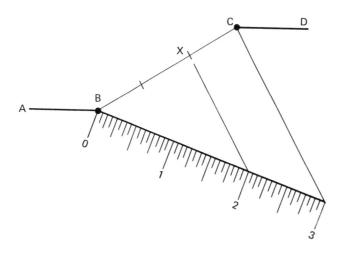

Step 2. Bisect line BX.

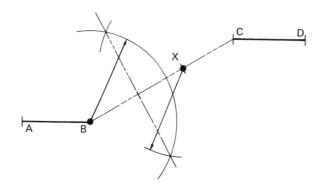

Step 3. Bisect line XC.

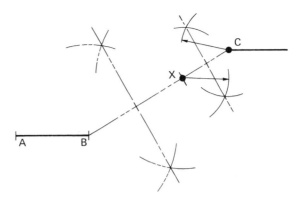

Step 4. Project 90 degrees upwards from point B and 90 degrees downwards from point C until they meet the bisectors of lines BX and XC. Swing points are located where these lines meet.

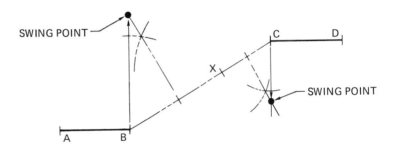

Step 5. Swing arcs BX and XC from swing points to complete drawing. Use hash marks to indicate tangent points.

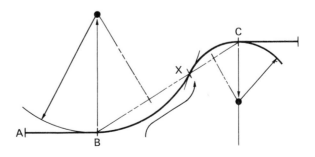

HOW TO CONSTRUCT AN ARC TANGENT TO TWO LINES AT RIGHT ANGLES (90°)

Given:

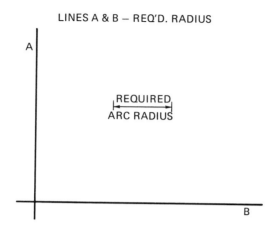

LINES A & B – REQ'D. RADIUS

REQUIRED ARC RADIUS

Step 1. Out of the way, swing an arc from line A and another arc from line B that are both equal to the radius required.

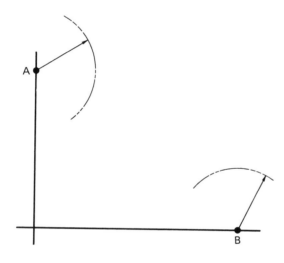

Step 2. Draw lines tangent to arcs A and B and parallel to lines A and B.

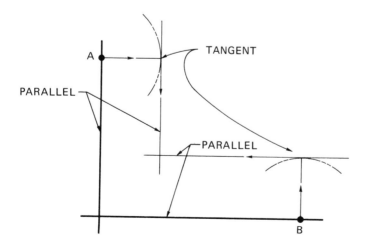

Step 3. Where these lines cross is the center for the required radius. Swing the required radius. Add tangent points and thicken to correct line weight.

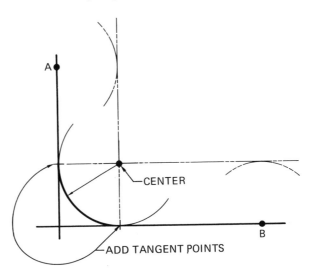

HOW TO CONSTRUCT AN ARC TANGENT TO AN OBTUSE ANGLE (MORE THAN 90°)

Given:

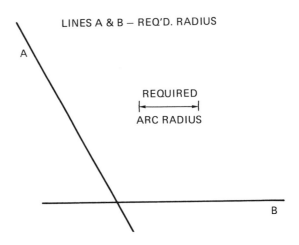

Step 1. Out of the way, swing an arc from line A and another arc from line B that are both equal to the radius required.

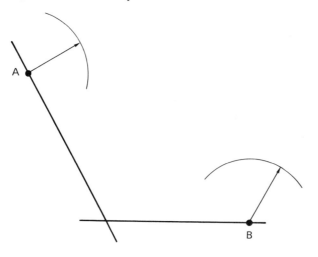

Step 2. Draw lines tangent to arcs A and B and parallel to lines A and B.

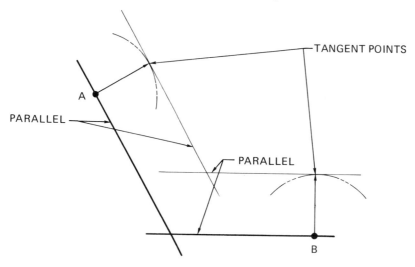

Step 3. Where these lines cross is the center for the required radius. Swing the required radius. Add tangent points and thicken to correct line weight.

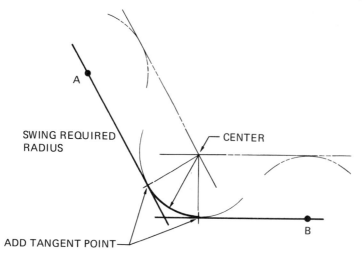

**HOW TO CONSTRUCT AN ARC TANGENT TO
AN ACUTE ANGLE (LESS THAN 90°)**

Given: LINES A & B — REQUIRED RADIUS

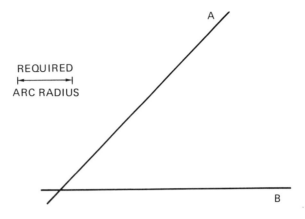

Step 1. Out of the way, swing an arc from lines A and B that are both equal to the radius required.

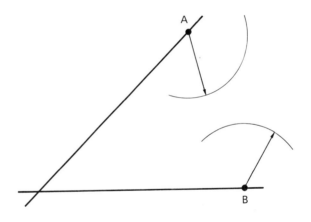

Step 2. Draw lines tangent to arcs A and B and parallel to lines A and B.

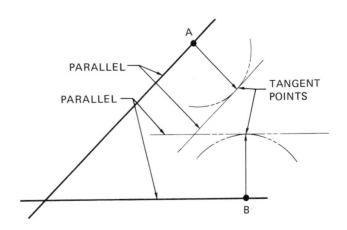

Step 3. Where these lines cross is the center for the required radius. Swing the required radius. Add tangent points and thicken to correct line weight.

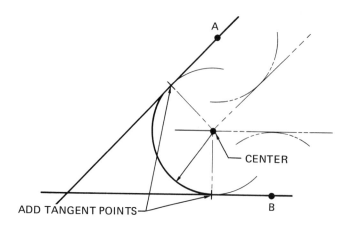

HOW TO CONSTRUCT AN ARC TANGENT TO A STRAIGHT AND A CURVED LINE

Given: LINES A AND B — REQUIRED RADIUS

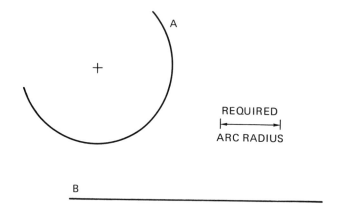

Step 1. Out of the way, swing an arc from line A and another arc from line B that are both equal to the radius required.

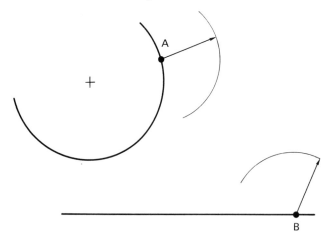

Step 2. Draw lines tangent and parallel to lines A and B.

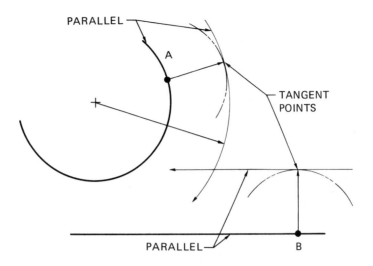

Step 3. Where these lines cross is the center for the required radius. Swing the required radius. Add tangent points and thicken to correct line weight.

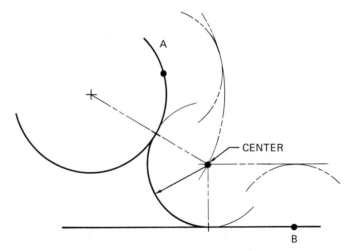

HOW TO CONSTRUCT AN ARC TANGENT TO TWO CURVED LINES

Given:

LINES A AND B

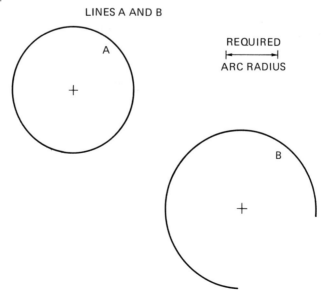

Step 1. Out of the way, swing an arc from line A and another from line B that are both equal to the radius required.

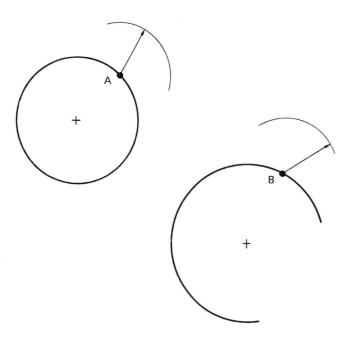

Step 2. Draw a line tangent and parallel to line A.

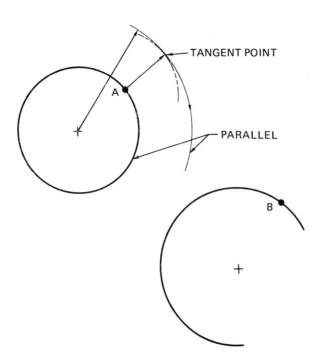

Step 3. Draw a line tangent and parallel to line B.

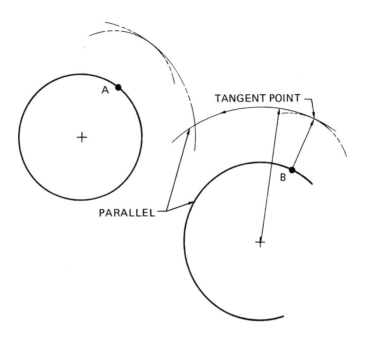

Step 4. Where these lines cross is the center for the required radius. Swing the required radius. Add tangent points and thicken to correct weight.

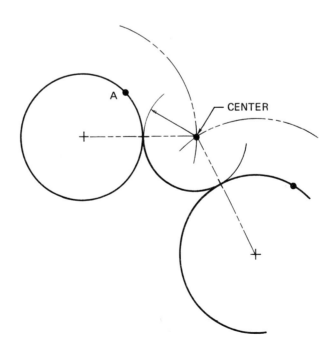

UNIT 5

MULTIVIEW DRAWINGS

MULTIVIEW DRAWING SYSTEM

A *multiview drawing* views an object from more than one place. The top view, for instance, shows how the object appears when looking directly down on it; the front view drawing shows how the object appears when looking directly at the front of it, and so on. A standard method of layout is followed so that all drawings are made in basically the same manner. Each view is in direct relationship to the next, but each view illustrates the object from a different viewpoint.

Figure 5-1 shows a pictorial view of a die to help explain the relationship between views. A *pictorial view* shows the object as it appears to the eye. Figure 5-2 shows how the die appears if viewed downward from above the die. Figure 5-3 shows how it appears looking at it directly from the front. Figure 5-4 shows the die viewed from the right side.

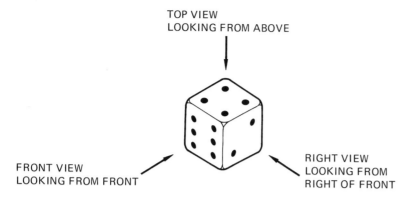

Fig. 5-1 Pictorial view of die

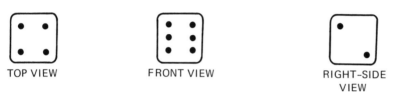

| TOP VIEW | FRONT VIEW | RIGHT-SIDE VIEW |

Fig. 5-2 Top view   Fig. 5-3 Front view   Fig. 5-4 Right-side view

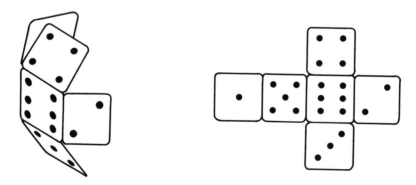

Fig. 5-5 Die opened up and flattened out

If the die is opened up and flattened out, figure 5-5, a better understanding of the multiviews can be gained.

Figure 5-6 shows the six sides of the die unattached to each other and illustrates the three most frequently used views in the circle: the top, front, and side views. Nearly all multiview drawings made of three views use those circled in figure 5-6.

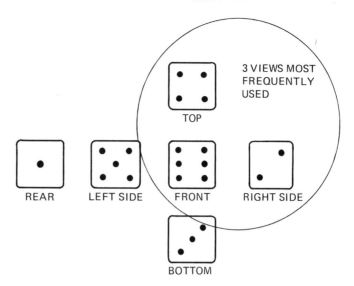

Fig. 5-6 Six sides of die, unattached

PLACEMENT OF VIEWS

Referring to the die, the side with six dots has the most detail. Use this side for the front view, figure 5-7. Directly to the right of the front view is the right-side view. It is projected from the front view as shown in figure 5-8. Directly above the front view is the top view. It, too, is projected from the front view as shown in figure 5-9.

These are the usual three views. If a left-side view is needed, it is projected from the front view to the left. The front, top, and right-side views are the views most often used.

Fig. 5-7 Front view

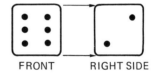

FRONT   RIGHT SIDE

Fig. 5-8 Project the right-side view from front view without measuring height.

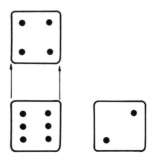

Fig. 5-9 Project the top view upwards from the front view without measuring the width.

ADDING 45-DEGREE ANGLE PROJECTION LINES (MITER LINES)

From surface A in the top view, project horizontally as shown in figure 5-10. From surface A in the right-side view, project vertically. Where projections cross, construct a 45-degree angle. Project all points up to the 45-degree line, called a *miter line,* and over to check work and save drawing time. This is done lightly with a 4H lead.

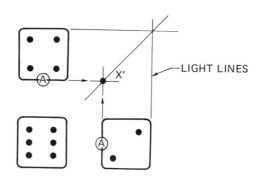

Fig. 5-10 45-degree projection lines

REQUIRED VIEWS

A drawing must be clean and simple so that there are no questions concerning what the drawing means.

The round cannon ball in figure 5-11 needs only one view to describe its size and shape. If the top and side views were added, it would be repetitious of the front view without providing any additional information.

An object such as the flat gasket in figure 5-12 needs only one view and a callout stating required thickness.

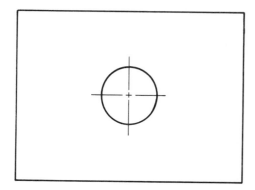 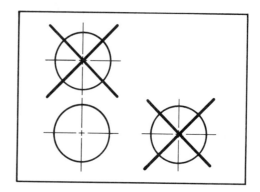

Fig. 5-11 One-view drawing

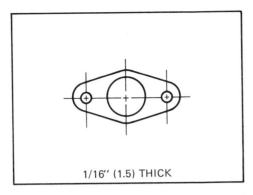 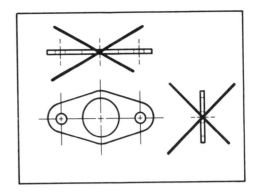

Fig. 5-12 One-view drawing

Many objects can be illustrated with only two views. A third only duplicates the same information. Figure 5-13 shows an example of a drawing requiring only two views. Do not use more views than needed to illustrate an object.

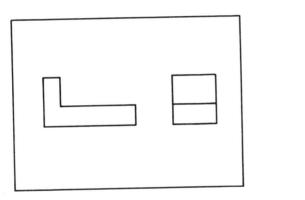 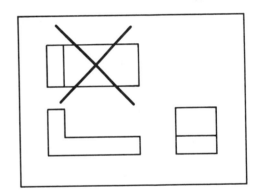

Fig. 5-13 Two-view drawing

In the simple system of multiview drawing, the front view is the starting view from which other views are made. The front view is first selected according to these rules:

- The front view is the most important view.

- The front view should show the basic shape in profile.

- The front view should be drawn so that it appears stable. To accomplish this, the heavy part is placed at the bottom of the view.

- The front view should be placed so that the other views have as few hidden edges as possible.

- The front view should show the most detail.

Before beginning a three-view drawing of any object, make some isometric sketches, such as the one illustrated in figure 5-14. Isometric drawing is discussed in Unit 6. This eliminates possible errors in the finished drawing and is helpful in selecting the proper view as discussed in rules for selecting a front view on pages 112 and 113.

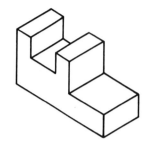

Fig. 5-14 Isometric of object

Follow these basic steps before beginning a three-view drawing:

1. Visualize the object.
2. Decide which view to use as the front view by sketching it in various positions, figure 5-15. Choose the one that most closely follows the rules for selecting the front views discussed on pages 112 and 113.

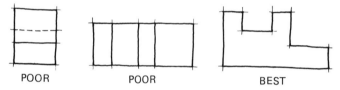

Fig. 5-15 Selecting proper front view

3. Decide how many views are needed to completely illustrate the object.
4. Decide in which position the front view will be placed. Figure 5-16 illustrates poor and good positioning.
5. Sketch all views. Use 1 inch (25) between views regardless of the scale being used. Make sketches freehand. Do not measure distances nor use a straightedge while sketching. Make sure views are neat and centered on the page, figure 5-17.

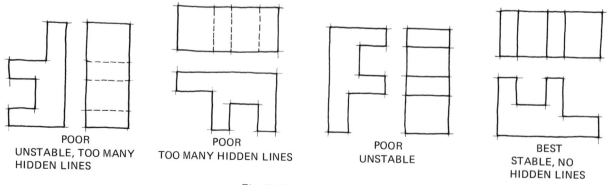

Fig. 5-16 Front view positioning

The example shown on the left of figure 5-17 is poorly centered. It has wasted space on both sides of the top and front views, and both views are too close to the top and bottom of the paper. Objects should never be drawn within 1/2 inch (12) of the border lines. The example shown on the right of figure 5-17 is better than the first example because it appears balanced and well centered.

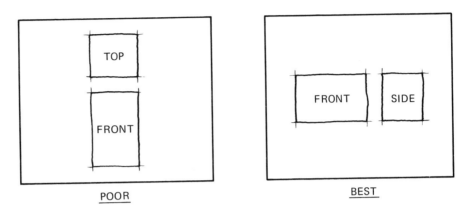

POOR   BEST

Fig. 5-17 Centering sketches

PRECEDENCE OF LINES

Sometimes lines coincide, figure 5-18. When this happens the drafter must choose which line to show. The order of precedence, or importance, of lines is:

1. Visible edge or object lines
2. Hidden lines
3. Center lines

Object lines, therefore, are more important than hidden lines. Hidden lines are more important than center lines. If a center line coincides with a cutting-plane line, the cutting-plane line is shown.

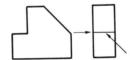

THE OBJECT LINE COINCIDES WITH THE HIDDEN LINE

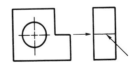

THE OBJECT LINE COINCIDES WITH THE CENTER LINE

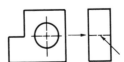

THE HIDDEN LINE COINCIDES WITH THE CENTER LINE

Fig. 5-18 Precedence of lines

CENTERING

The drawing must be centered within the work area of the paper or the border, if one is provided. A full 1 inch (25), regardless of the scale being used, should be placed between all views drawn in this text.

Figure 5-19 shows the procedure used to center a drawing within a specified work area.

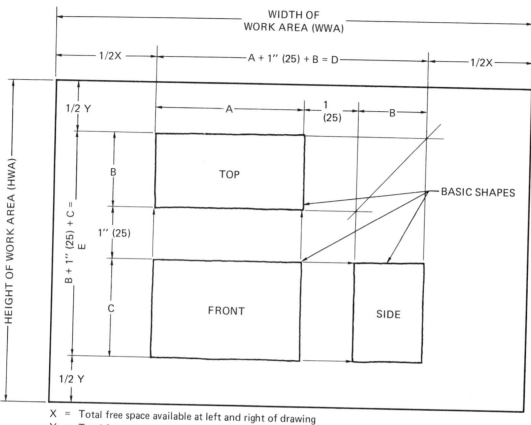

X = Total free space available at left and right of drawing
Y = Total free space available at top and bottom of drawing
D = Width of entire drawing, including 1" (25) between views
E = Height of entire drawing, including 1" (25) between views

Fig. 5-19 Centering sketch

The total width (D) of the drawing is determined by adding measurement A to measurement B and then adding 1 inch (25) for the distance between views. The total height (E) of the drawing is determined by adding B to C and then adding 1 inch (25) for the distance between views. The total height and width of the work area is found by measuring it directly from the paper.

To center the drawing horizontally, subtract D from the width of the work area. The answer represents the available free space. This free space (X) is divided by two. One half of X is placed to the right of the views, and one half of X is placed to the left of the views.

To center the drawing vertically, the same basic procedure is followed. Subtract E from the height of the work area, divide the answer (Y) by two, and place half of Y to the top of the drawing and half of Y to the bottom.

This centers the drawing within the work area. Regardless of the drawing size or available work area, the same process is followed each time a drawing is to be centered.

INTERSECTIONS

Figure 5-20 illustrates the basic methods of showing intersections on a drawing. When two lines intersect, the drawing must show if they touch or pass by each other. Note how these lines are drawn in the figure.

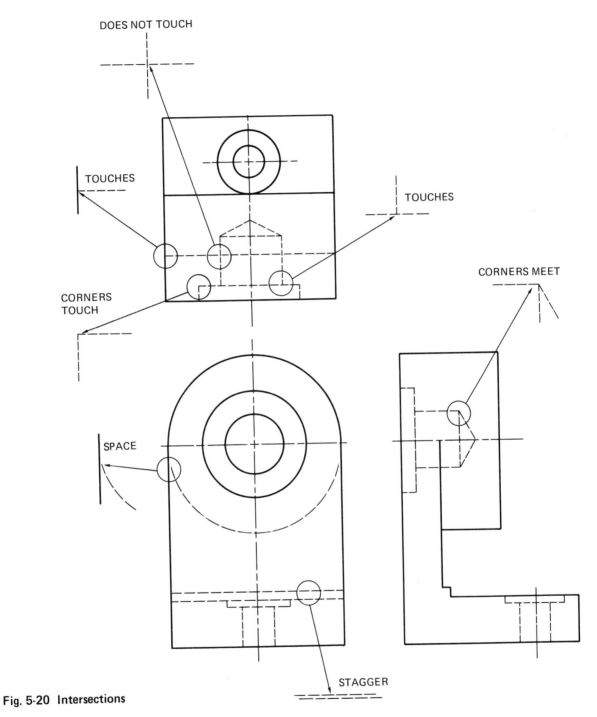

Fig. 5-20 Intersections

Hidden lines are drawn medium weight, figure 5-21. Note the size of the dash and space in the illustration.

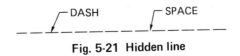

Fig. 5-21 Hidden line

NUMBERING DRAWINGS

Numbering each corner of the more difficult drawings makes them easier to visualize and draw. Think of a drawing as a group of points joined together by lines. If the ends of the lines can be found, all that has to be done to complete a drawing is to connect the ends. Once the ends have been found and numbered in one view, the same can be done in other views by projection, figure 5-22.

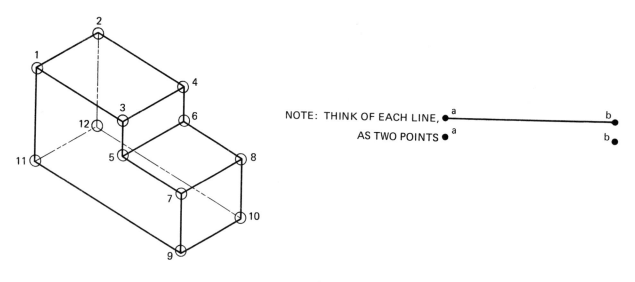

NOTE: THINK OF EACH LINE, ●———●
AS TWO POINTS ●ᵃ   b●

Fig. 5-22 An isometric drawing with each corner numbered and ends connected.

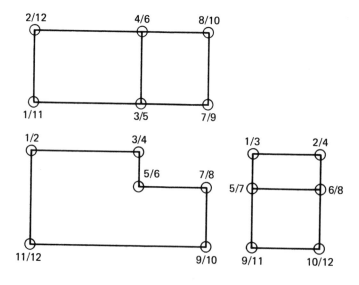

A three-view drawing of the isometric drawing above (Fig. 5-22) with each corner numbered and ends connected.

PROJECTING NUMBERS

Given:

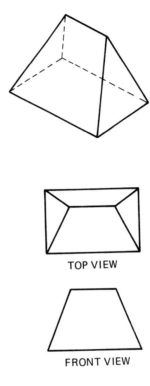

TOP VIEW

FRONT VIEW

Fig. 5-23

Step 1. Assign a number to each corner of the top view of figure 5-23. Project those points down to the front view. Number the points in the front view.

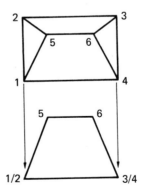

Step 2. Project the points across from the front view, and the same points from the top view, over to the 45-degree line. Where the lines meet the 45-degree line, project down to the side view. The points are located where these lines cross. Number all points.

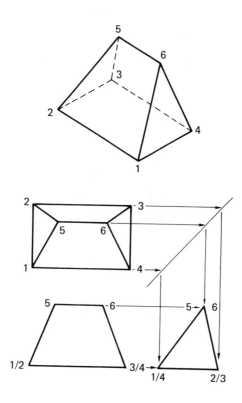

ROUNDS AND FILLETS

Rounds are used to remove the sharp edges on cast objects by rounding outside corners.

Fillets are inside rounded corners. Cast objects tend to crack due to the strain placed on the metal during the cooling process. Fillets distribute the strain and prevent cracking.

Rounds and fillets also enhance the appearance of an object. Examples of rounds and fillets are shown in figure 5-24 and 5-25.

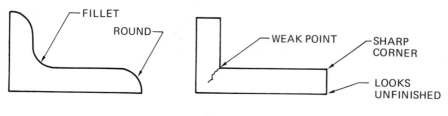

Fig. 5-24

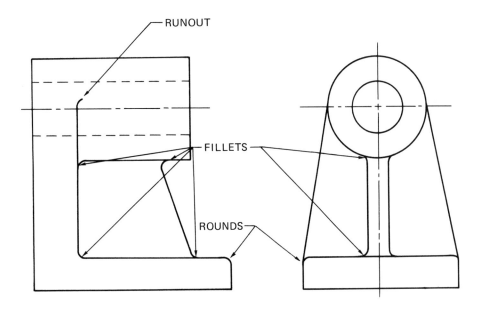

Fig. 5-25 Examples of runouts, fillets, and rounds

RUNOUTS

Runouts are curved surfaces formed where a flat and curved surface meet.

Using figure 5-26, follow these steps to find the intersection where a runout will be drawn.

1. Locate the tangent points.
2. Project the tangent points to the next view.
3. Add the runouts as shown.

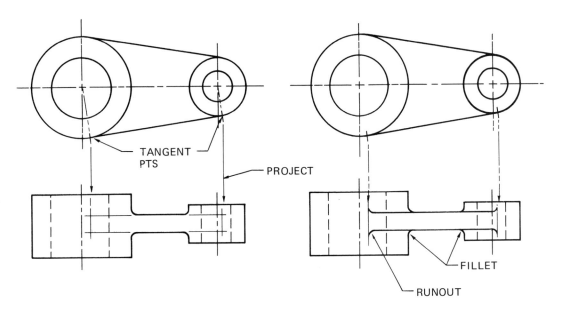

Fig. 5-26

TREATMENT OF INTERSECTING SURFACES

Figures 5-27 through 5-30 show how to draw intersecting surfaces. In figure 5-27, the flat surface meets the cylinder sharply, making a visible edge necessary in the front view.

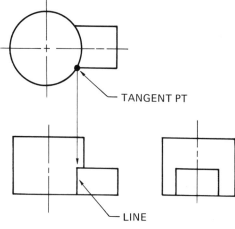

Fig. 5-27

In figure 5-28, the flat surface blends into the cylinder, stopping at the point of tangency.

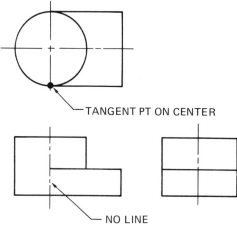

Fig. 5-28

In figure 5-29, the surface blends into the cylinder, stopping at the point of tangency.

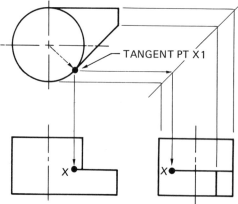

Fig. 5-29

In figure 5-30, the radius blends into the base, stopping at the point of tangency.

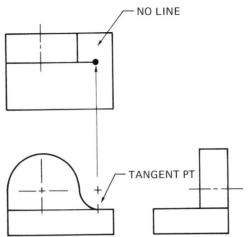

Fig. 5-30

CURVE PLOTTING

Using figure 5-31, follow the steps in curve plotting.

Step 1. Lightly complete the basic shape of each view.

Step 2. Locate the center of the arc, point "x."

Step 3. From this point, divide the arc into equal spaces. Figure 5-31 uses 30-degree spaces, but the more spaces, the more accurate the drawing.

Step 4. Number the points in order, clockwise.

Step 5. Project these points up from the right-side view to the 45-degree axis line and to the left.

Step 6. Project over to the slanted surface in the front view and then upwards.

Step 7. Where the lines projected from the given points in the side and front views meet will form the points 1 through 7 in the top view. Number all points at their intersections.

Step 8. Using a French curve, complete the drawing.

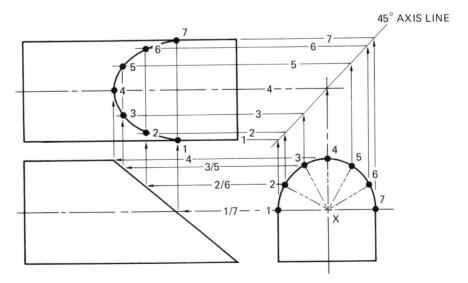

Fig. 5-31

UNIT 6

BASIC ISOMETRICS

ISOMETRIC

An *isometric drawing* is a form of pictorial drawing used to make three-view drawings easier to visualize. It is made with a 30°–60° triangle. An isometric is actually a three-dimensional drawing because height, width, and depth are all drawn on a single plane. Figure 6-1 illustrates the difference between an isometric and orthographic view. An *orthographic view* is formed by projecting perpendicular lines from an object onto a single plane.

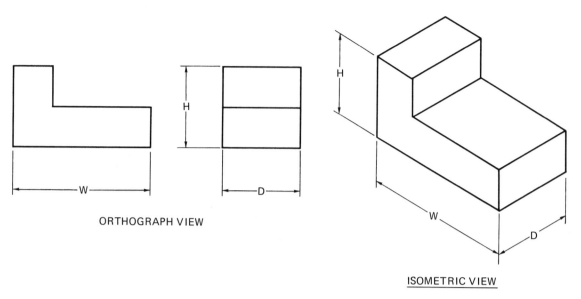

ORTHOGRAPH VIEW

ISOMETRIC VIEW

Fig. 6-1 Orthographic and isometric drawings of the same object.

Isometric lines are drawn parallel to an isometric axis, figure 6-2. Horizontal and vertical measurements are taken directly from the orthographic view and transferred to the corresponding isometric lines. Measurements may be made on any isometric axis line or lines parallel to them. Hidden edges are not shown on isometric drawings.

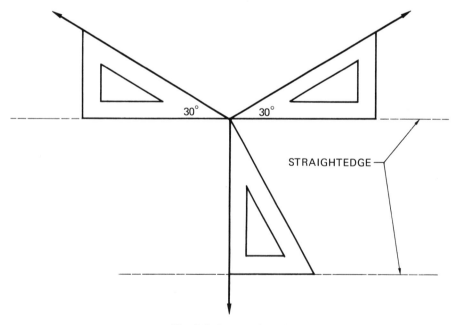

Fig. 6-2 Isometric axis

NON-ISOMETRIC LINES

A *non-isometric line* is one that is not parallel to any of the lines on an isometric axis. Angles, for instance, are non-isometric lines and cannot be found with a protractor. Non-isometric lines are found by locating their ends with points and then connecting these points.

Plotting Non-Isometric Lines

Steps 1 through 4 outline the procedure to follow to plot non-isometric lines.

Step 1. Number the ends of non-isometric lines.

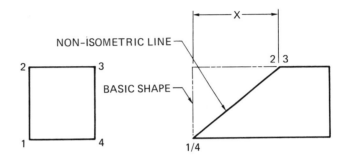

Step 2. Lightly draw the basic shape in isometric.

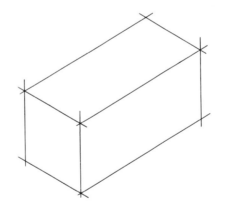

Step 3. Locate points or *ends of lines*. Note measurement "x."

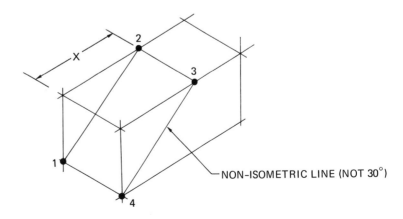

Step 4. Complete the drawing with all lines of object line thickness.

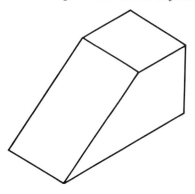

HIDDEN EDGES IN ISOMETRIC

Hidden edges are shown in an isometric drawing only if it is the only drawing available for measurement and detail. Isometrics, however, are most often shown along with a three-view drawing for clarification. In such cases, hidden edge lines are not used.

ISOMETRIC CONSTRUCTION

Box Method

Figure 6-3 shows one method of isometric construction when drawing an object made up of non-isometric lines. In this case a box is made using the basic shape and size of the figure. Known points are plotted on the isometric lines and joined to form the completed isometric drawing.

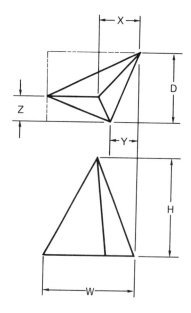

Fig. 6-3 Two-view drawing of a pyramid

Step 1. Lightly draw the basic shape using measurements taken directly from the two-view drawing.

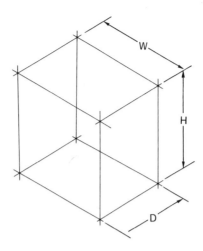

Step 2. Locate all points. Project at 30 degrees where necessary to locate positions which are not on the box lines.

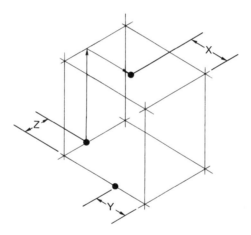

Step 3. Using the two-view drawing as a guide, complete the drawing by connecting all plotted points.

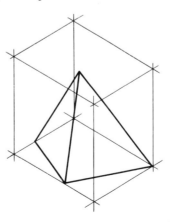

ISOMETRIC CONSTRUCTION

Skeleton Method

The skeleton method of isometric drawing is very similar to the box method. The main difference between the two is that only the base of the basic shape is drawn when using the skeleton method, and heights are located by drawing vertical lines from the points located on the base. The top of the pyramid shown in figure 6-4 is located H distance above the point at which X and Z intersect.

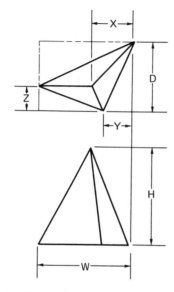

Fig. 6-4 Two-view drawing of a pyramid

Step 1. Lightly draw the base using measurements taken directly from the two-view drawing.

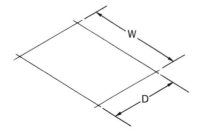

Step 2. Locate all points on the base and project them at 30 degrees to find height position. Draw a vertical line where lines X and Z cross. Locate the height (H) on this line.

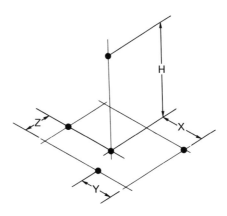

Step 3. Using the two-view drawing as a guide, complete the drawing by connecting all plotted points.

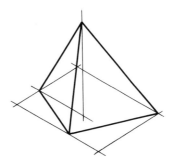

CURVES IN ISOMETRIC

Lines which make up curves in isometric drawings are non-isometric lines and are made by plotting points, figure 6-5.

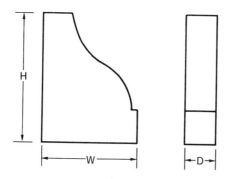

Fig. 6-5 Two-view drawing

Step 1. Draw several vertical or horizontal lines on the surface of the view. Number the ends of these lines.

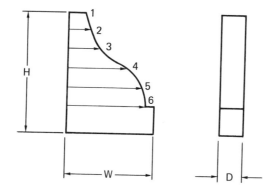

Step 2. Lightly draw the basic shape using the box construction method. Draw lines 1 through 6 at 30-degree spaces as in step 1. Transfer line lengths 1 through 6. Lay out the curve with a French curve.

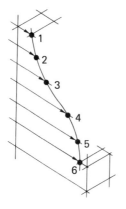

Step 3. Project lines 1 through 6 at 30-degree to the right and measure the thickness on each line. Draw the opposite curve with a French curve.

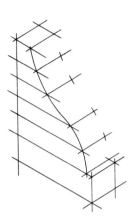

Step 4. Darken all lines using correct line weight.

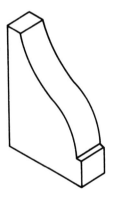

CENTERING AN ISOMETRIC DRAWING

To center an isometric drawing in a given work area, follow steps 1 through 5.

Step 1. Locate the center of the work area by drawing light diagonal lines from corner to corner. Where these lines intersect form the center of the work area.

Step 2. Draw a light line downward from the center of the work area that measures half the height (H) of the basic shape.

Step 3. Draw a 30-degree line down to the left from the end of the line made in step 2 that measures half the depth (D) of the basic shape.

Step 4. Draw a 30-degree line down to the right from the end of the line made in step 3 that measures half the length (L) of the basic shape.

Step 5. From the end of the line formed in step 4, draw 30-degree lines upward to the left and right, and a vertical line upwards. Using full size measurements taken from the given view in figure 6-6, complete the basic shape in the work area.

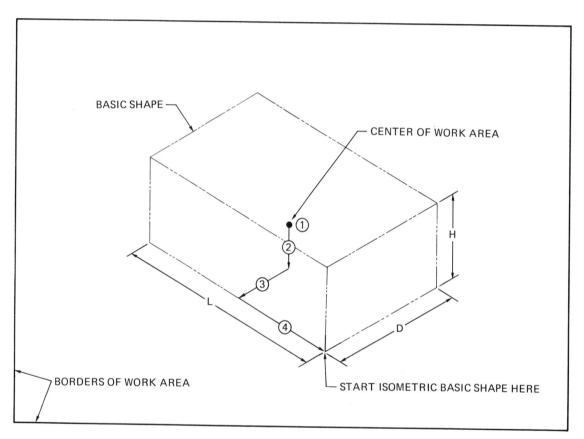

Fig. 6-6 Centering an isometric drawing

ISOMETRIC CIRCLES

There are two methods used to make circles in isometric. Method 1 is faster to use than method 2 but not as accurate.

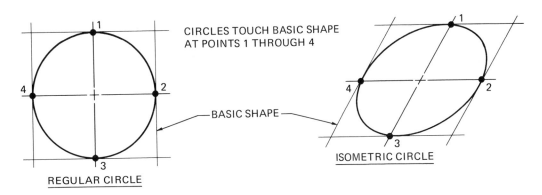

Fig. 6-7 Regular and isometric circles

Method 1

Step 1. Draw a horizontal center line. Through the center, draw an isometric vertical center line using a 60-degree angle. Set the compass at the radius required and swing the arcs to cut center lines.

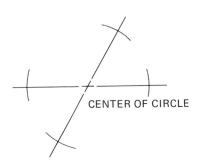

Step 2. Draw the basic shape and locate points 1 through 4. Draw perpendicular lines to the basic shape from the key points.

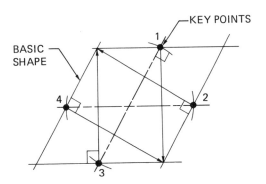

Step 3. Center points for radii 1–2 and 3–4 are located where perpendicular lines cross at points a and b. Swing arcs 1–2 and 3–4.

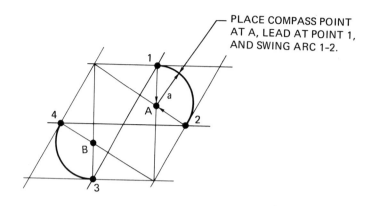

PLACE COMPASS POINT AT A, LEAD AT POINT 1, AND SWING ARC 1-2.

Step 4. Center points for radii 2–3 and 1–4 are found at outside corners c and d. Swing arcs 2–3 and 1–4. Each radius forms 1/4 of an isometric circle. Arcs, therefore, are drawn from one center line to the next center line. Darken all arcs using correct line weight.

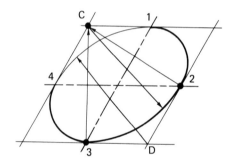

ISOMETRIC RADII

Isometric radii are drawn using the same procedure as method 1 for drawing isometric circles except that only that part of the isometric basic shape needed is used.

In figure 6-8, point A is used as the center for arc 1-2. The inside arc shows the thickness and is drawn from point A′. Point A′ is formed by drawing a 30-degree line from point A that measures the width of the stock given. The radius for the new arc will be exactly the same as arc 1-2. The same procedure is used to locate point C′. Point B, the center for radius 2-3, is found using the procedure detailed in method 1 for drawing isometric circles.

BASIC
SHAPE

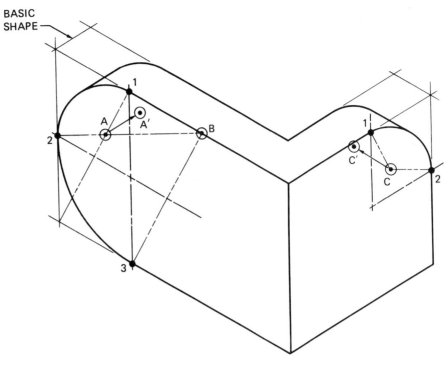

Fig. 6-8

ISOMETRIC CIRCLES

Method 2

The procedure for making isometric circles using method 2 is slower than method 1, page 81, because it is done by plotting points, then connecting those points with French curves. It is, however, more accurate than method 1, figure 6-9.

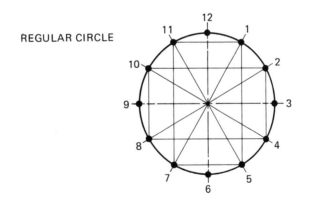

REGULAR CIRCLE

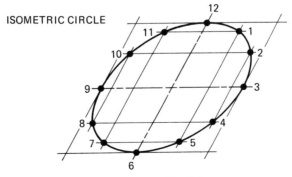

ISOMETRIC CIRCLE

Fig. 6-9

Step 1. On scrap paper, draw a full-size circle using the diameter of the given object. Divide the circle into twelve equal parts with a 30°–60° triangle. Number all parts.

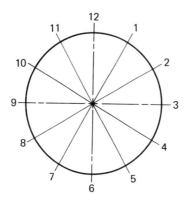

Step 2. Where the lines cross the circle draw rectangles.

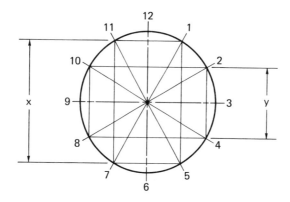

Step 3. Begin the finished drawing by locating the center lines of the isometric circle. With dividers, locate points 3, 6, 9, and 12 on the isometric center lines. Complete the basic shape.

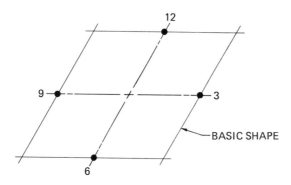

Step 4. From the sketch made in steps 1 and 2, transfer measurements on and about isometric center lines. Number all points. Connect with French curves. Darken drawing to correct line weight.

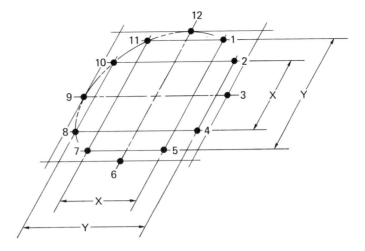

ISOMETRIC ELLIPSE

An ellipse can be drawn in isometric in much the same way as isometric circles by using method 2.

A truncated cylinder is illustrated in figure 6-10. The word *truncated* means cut off. The basic principle of this method is to consider the various points in the basic shape as having height. If all heights were drawn from the points in the basic shape, and all heights were located on these straws, the points forming the ellipse would be located. These points would be connected with French curves to form the ellipse.

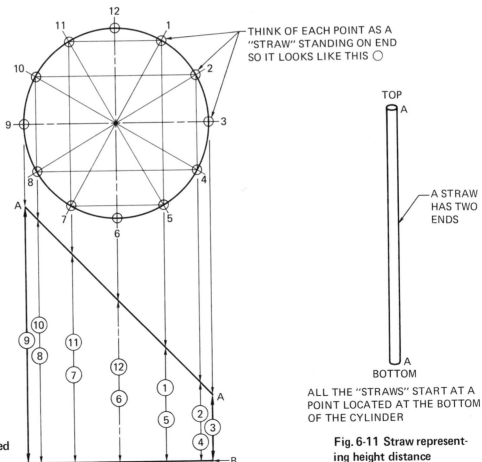

Fig. 6-10 Truncated cylinder

Fig. 6-11 Straw representing height distance

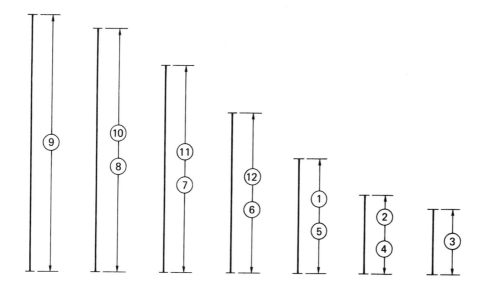

Fig. 6-12 Examples of line lengths taken from figure 6-10

How to Draw an Ellipse

Step 1. Draw basic shape of cylinder shown in figure 6-10.

Step 2. Draw an isometric circle within basic shape of cylinder base, using method 2.

Step 3. Add points 1 through 12.

Step 4. Project straws 30 degrees from each point. On these straws, transfer lengths of lines from the given view shown in figure 6-10.

Step 5. Connect the ends of these lines with French curves.

Step 6. Thicken lines.

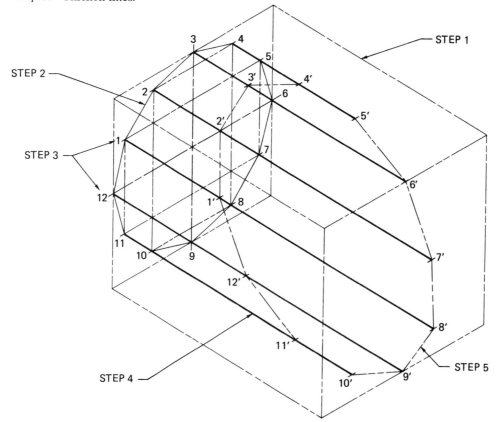

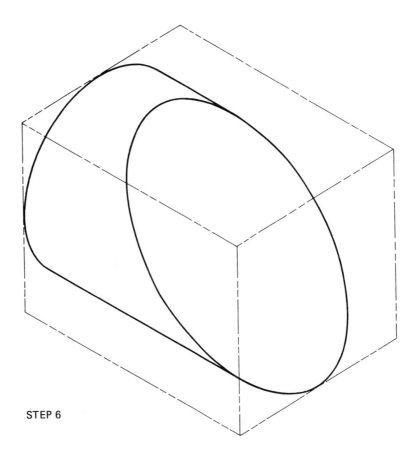

STEP 6

SHADING

The main purpose of a pictorial drawing is to illustrate the object as close to what it actually looks like as possible. Shading gives the illusion of distance and depth, creating a three-dimensional effect on a flat sheet of paper.

Only three sides of an object are normally shown in a pictorial drawing. The horizontal surface is usually shaded the lightest in value, the frontal plane is a medium value overall, and the profile plane is shaded the darkest, figure 6-13. An edge is formed where contrasting values meet.

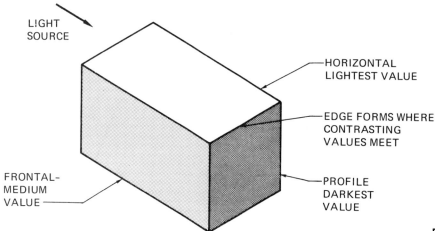

LIGHT SOURCE

HORIZONTAL LIGHTEST VALUE

EDGE FORMS WHERE CONTRASTING VALUES MEET

FRONTAL-MEDIUM VALUE

PROFILE DARKEST VALUE

Fig. 6-13 Shading

Line Shading

Line shading varies the light intensity of the paper surface by placing thin lines closer together.

First, think of where the light is coming from. It is usually from over the left shoulder and from above. Thus, where the light shines is the lightest. It gets progressively darker as movement away from the light and into the shadows takes place, figure 6-14.

In line shading, simply draw the lines closer together for a darker effect and space them out further as the light source is approached, figure 6-15.

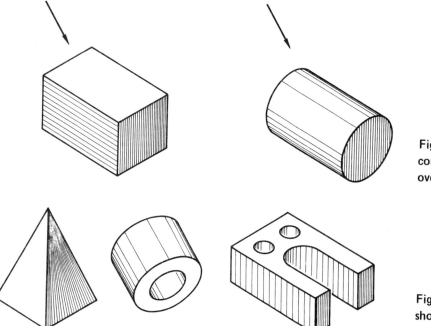

Fig. 6-14 Light source coming from above and over the left shoulder

Fig. 6-15 Shading lines should be drawn parallel to the edge of the surface.

Stippling

Stippling is a method of shading that uses a series of dots placed close together to represent dark areas and spaced apart to represent light areas. The same basic instructions for line shading apply to stippling. Stippling produces pleasing and realistic illustrations, figures 6-16 and 6-17.

There are *shading films* available on the market which can eliminate the repetitious job of stippling. In order to use these films most effectively, other methods of shading must first be mastered.

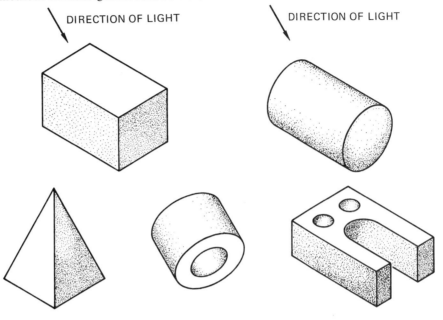

Fig. 6-16 The light source is shown originating above and from the left.

Fig. 6-17 Examples of objects which have been shaded using the stippling method

Pencil Shading

Pencil shading uses the same theory as line shading and stippling. A 2H or softer pencil is used to pencil shade.

Loose *graphite* is often used to improve the appearance of drawings shaded with pencils. The graphite is placed on the drawing and smudged with the index finger or a tightly rolled piece of paper. All changes should be gradual from white to very black. An eraser and erasing shield is used to erase smudges that go outside of the drawing area and to highlight any important light areas. Practice this method on scrap paper, figure 6-18.

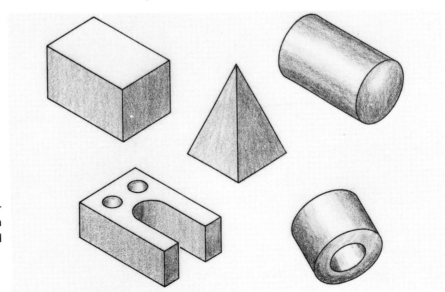

Fig. 6-18 Examples of objects which have been shaded with pencil and graphite

Adding Shadows

Sometimes a shadow is required in order to further emphasize the object. Figure 6-19 illustrates a suggested method of adding shadows to finished objects.

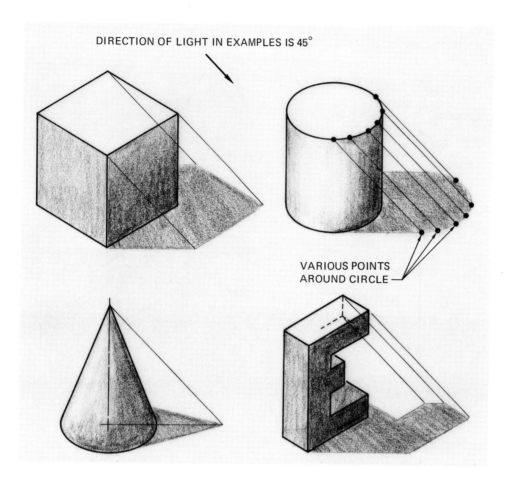

DIRECTION OF LIGHT IN EXAMPLES IS 45°

VARIOUS POINTS
AROUND CIRCLE

Fig. 6-19 Adding shadows

UNIT 7

SECTION
VIEWS

Usually a drawing is represented by three regular views — front, top, and right side, unless the object is simple and three views are repetitive. All hidden surfaces are shown by hidden lines. Sometimes, however, there are so many hidden lines in a view it is impossible to visualize the object. In these cases, a section view is drawn.

Cutting-Plane Line

Cutting-plane lines are imaginary cuts through the object. Think of a saw cutting through an object. The cutting-plane line is represented by a thick, dark line, figure 7-1.

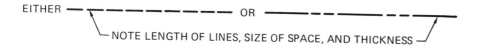

Fig. 7-1 Two methods of drawing cutting-plane lines

To better visualize this, figure 7-2 shows an imaginary cutting plane passing through an object. In figure 7-3, the object has been cut in half. The arrow shows the surface which will be drawn as a cut section drawing. The completed cut section drawing is shown in figure 7-4.

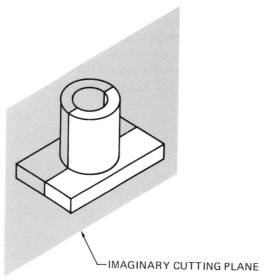

IMAGINARY CUTTING PLANE

Fig. 7-2

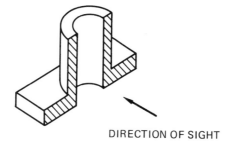

DIRECTION OF SIGHT

Fig. 7-3

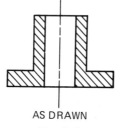

AS DRAWN

Fig. 7-4

Figure 7-5 (a) illustrates the object with an imaginary cutting plane passing through it. Figure 7-5 (b) shows that the cutting-plane line is drawn through the object. In this case, the plane cuts downward from the top view so the cutting-plane line is shown in the top view. Arrows are placed at the ends of the cutting line and are faced in the direction of sight. What remains after the object has been cut is projected into the front view and shown sectioned, as illustrated in figure 7-5 (c).

In figure 7-5 (c), the cutting plane is passed through from front to back. The bottom piece remains. The cut section line, therefore, is located in the front view, the direction of sight is downward, and the top view is shown as the sectioned view.

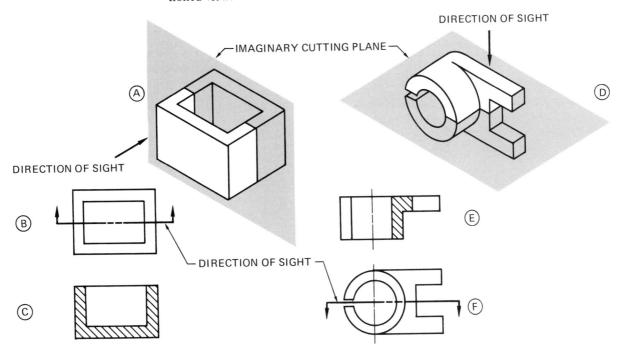

Fig. 7-5 **Cut-section drawing examples**

Section Lining

Section lining shows where the cutting plane passes, revealing the material used to make the object. Section lines are dark, thin lines drawn at a 45-degree angle and uniformly spaced by eye, figure 7-6. Though there are various symbols to represent specific kinds of materials, they are seldom used. Specific materials are identified by name on the detail drawing or parts list for the object.

If the cut object is made up of two pieces, both cut sections are drawn at a 45-degree angle, but in opposite directions, as shown in figure 7-7.

If the cut object is made up of three pieces, the cut section lines are drawn at 45-degree angles as shown in figure 7-8.

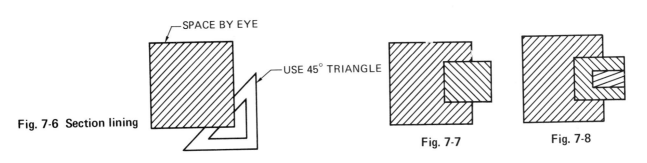

Fig. 7-6 **Section lining**

Fig. 7-7

Fig. 7-8

KINDS OF SECTION VIEWS

Full Section

A *full section* is simply a regular view that has been cut all the way through. All hidden lines can now be seen so they become visible lines. In figure 7-9, the hidden lines in the front view are not easy to visualize. A full section is therefore made of the object.

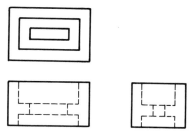

Fig. 7-9 Regular three-view drawing

Step 1. Add an imaginary cutting plane.

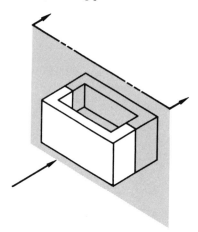

Step 2. Think of the front view as it would appear cut.

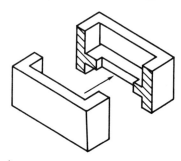

Step 3. Draw the views.

CUTTING-PLANE LINE. THINK OF IT AS A PLANE STANDING ON END.

ARROWS SHOW WHICH WAY THE DRAFTER IS VIEWING THE SECTION VIEW.

NO HIDDEN LINES IN SECTION VIEW

SECTION LINING IS DRAWN ONLY WHERE CUTTING PLANE CUTS THROUGH THE OBJECT.

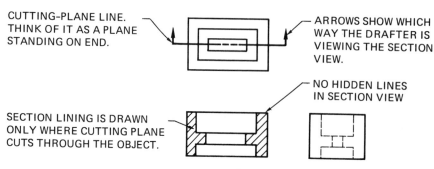

FULL SECTION

Half Section

In a *half-section drawing,* the object is cut only halfway through and a quarter is removed. A half section shows the inside of the object as well as the outside in the same view. This type of section is best used when the object is *symmetrical;* that is, the same shape and size on both sides of the center line. To make a half section drawing of figure 7-10:

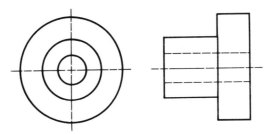

Fig. 7-10 Regular two-view drawing

Step 1. Add an imaginary cutting-plane line.

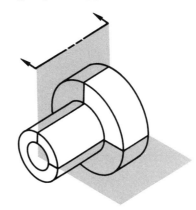

Step 2. Think of the right-side view as it would appear cut.

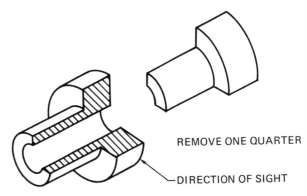

REMOVE ONE QUARTER

DIRECTION OF SIGHT

Step 3. Draw the views.

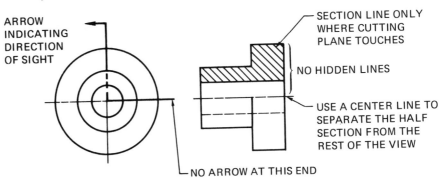

ARROW INDICATING DIRECTION OF SIGHT

SECTION LINE ONLY WHERE CUTTING PLANE TOUCHES

NO HIDDEN LINES

USE A CENTER LINE TO SEPARATE THE HALF SECTION FROM THE REST OF THE VIEW

NO ARROW AT THIS END

Broken-Out Section

Sometimes only a small area needs to be sectioned in order to make the interior of an object easy to understand. In this case a *broken-out section* is used. The broken line is put in freehand and is made visible-line thickness. The cutting-plane line can be omitted because it coincides with the center line. Figure 7-11 illustrates the procedure used to make a broken-out section drawing.

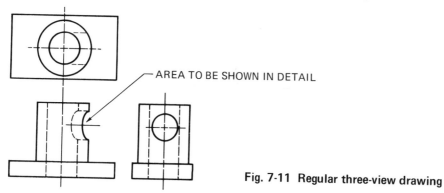

—AREA TO BE SHOWN IN DETAIL

Fig. 7-11 Regular three-view drawing

Step 1. Add an imaginary cutting plane through the area to be broken out.

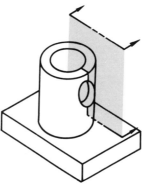

Step 2. Think of that area as it would appear broken out.

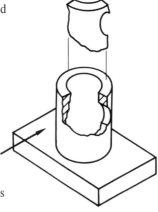

Step 3. Draw the views. No cutting lines are required.

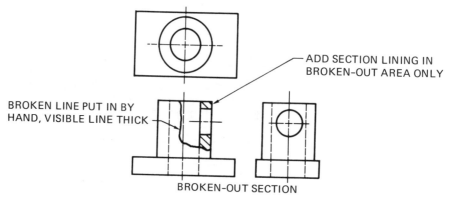

ADD SECTION LINING IN
BROKEN-OUT AREA ONLY

BROKEN LINE PUT IN BY
HAND, VISIBLE LINE THICK

BROKEN-OUT SECTION

Offset Section

An *offset section* is done as a full section and shows details that do not appear in a true full-section drawing. The cutting line is bent at 90 degrees to pick up important details. The bends of the cutting lines are not projected to or shown in other views. To make an offset section of figure 7-12:

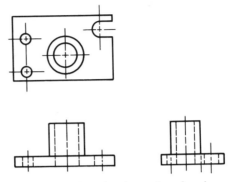

Fig. 7-12 Regular three-view drawing

Step 1. Add an imaginary cutting-plane line. Have it bend at 90-degree turns so it passes through the important features that must be described.

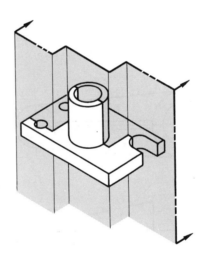

Step 2. Think of the view as it would appear cut.

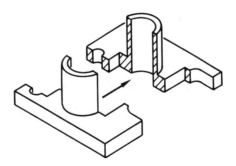

Step 3. Draw the views. The right side is not needed. Section only where the cutting-plane line cuts through the object.

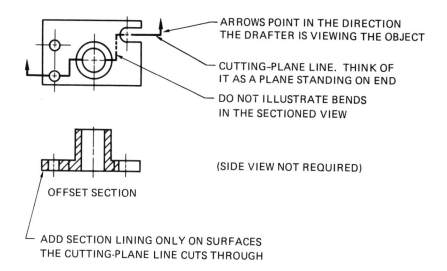

ARROWS POINT IN THE DIRECTION
THE DRAFTER IS VIEWING THE OBJECT

CUTTING-PLANE LINE. THINK OF
IT AS A PLANE STANDING ON END

DO NOT ILLUSTRATE BENDS
IN THE SECTIONED VIEW

(SIDE VIEW NOT REQUIRED)

OFFSET SECTION

ADD SECTION LINING ONLY ON SURFACES
THE CUTTING-PLANE LINE CUTS THROUGH

Revolved Sections

A *revolved section* shows the cross section of an object at a given location along its length. Revolved sections are cut perpendicular to the length of the part being shown. In figure 7-13, it is difficult to know what the cross section of the area indicated looks like.

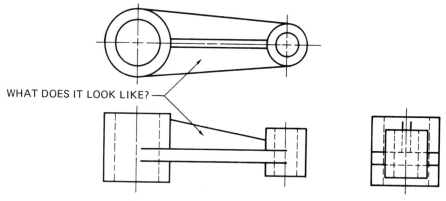

WHAT DOES IT LOOK LIKE?

Fig. 7-13 Regular three-view drawing

Step 1. Add an imaginary cutting-plane line through the area to be sectioned.

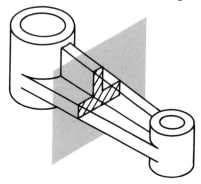

Step 2. Rotate the cutting-plane line, in place, and think of it as it would appear in the front view. Find the widths from other view, as shown.

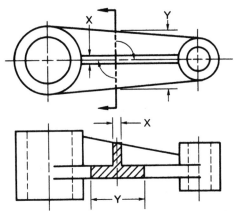

Step 3. Draw the views.

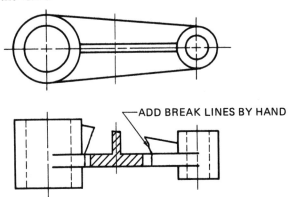

Removed Section

A *removed section* is the same as a revolved section except, as the name implies, the section is "removed" and drawn away from the object. Study the examples in figure 7-14. Note how they are called out; i.e., section A-A.

REVOLVED SECTION

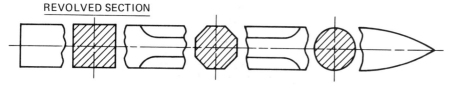

THIS IS HOW A REVOLVED SECTION WOULD LOOK. A REMOVED
SECTION IS SIMILAR TO A REVOLVED SECTION, BUT IT IS
SEPARATED FROM THE MAIN DRAWING. STUDY THE ILLUSTRATION
BELOW.

REMOVED SECTION

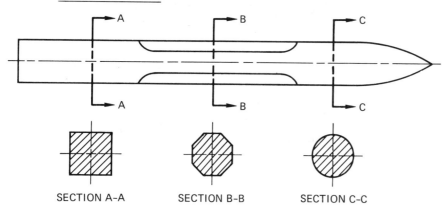

SECTION A-A   SECTION B-B   SECTION C-C

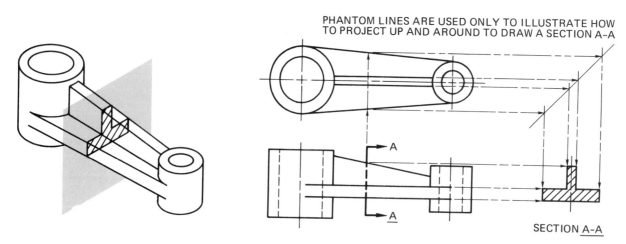

PHANTOM LINES ARE USED ONLY TO ILLUSTRATE HOW
TO PROJECT UP AND AROUND TO DRAW A SECTION A-A

SECTION A-A

FOLLOW THE SAME STEPS USED FOR A REVOLVED
SECTION EXCEPT REMOVE THE SECTION VIEW.

Fig. 7-14 Explanation for making a drawing with a removed section

Assembly Section

When a section drawing shows two or more parts together, it is called an *assembly drawing*, figure 7-15. Each part is called out with name, number, and number of parts required per assembly. These are identified on the parts list. Assembly sections can be drawn as full sections, half sections, or offset sections. The #1 callout identifies the assembly.

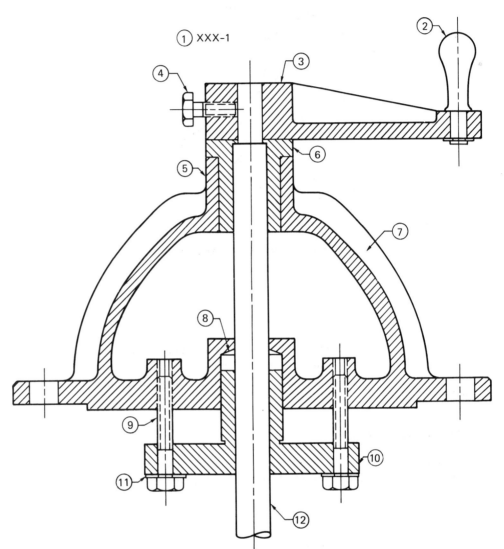

Fig. 7-15 Assembly drawing

FILLETS, ROUNDS, AND RUNOUTS

Most castings and forgings use fillets, rounds, and runouts, figure 7-16. These must be shown in section views. They are usually of an even radius, such as 1/16, 1/8, or 1/4.

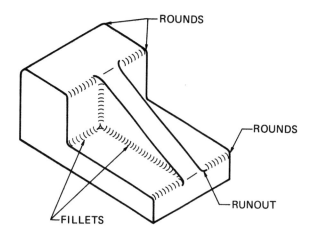

Fig. 7-16 Fillets, rounds, and runouts

THIN-WALL SECTION

Any object that is drawn in section and is very thin, such as a gasket or brass shim, should be filled in solid black as it is impossible to show the correct section lining. This is called a *thin-wall section*. Note the example of a thin-wall section in figure 7-17.

NONSECTIONED PARTS

If the cutting plane passes along the center of a bolt, screw, nut, pin, key, shaft, or rivet, it is not sectioned. Figure 7-17 illustrates a few examples of non-sectioned parts.

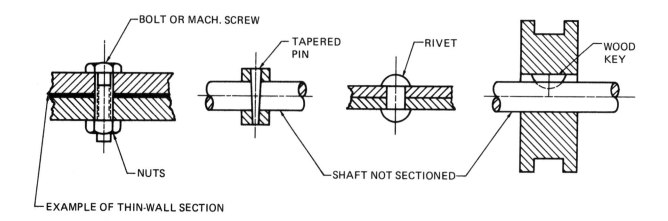

Fig. 7-17 Examples of nonsectioned parts

CONVENTIONAL BREAKS

Long objects, such as the pipe illustrated in figure 7-18, would appear very long if the entire length is drawn. Usually such objects are drawn to scale and shortened by using a *conventional break.*

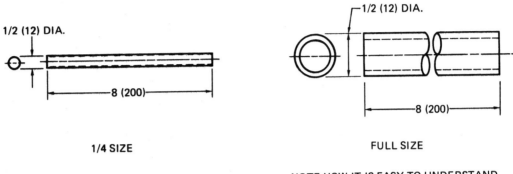

1/2 (12) DIA.

8 (200)

1/4 SIZE

NOTE HOW SMALL THE PIPE APPEARS.

1/2 (12) DIA.

8 (200)

FULL SIZE

NOTE HOW IT IS EASY TO UNDERSTAND

Fig. 7-18 Use of conventional break

Other Breaks

Study figure 7-19. It illustrates other methods of drawing breaks.

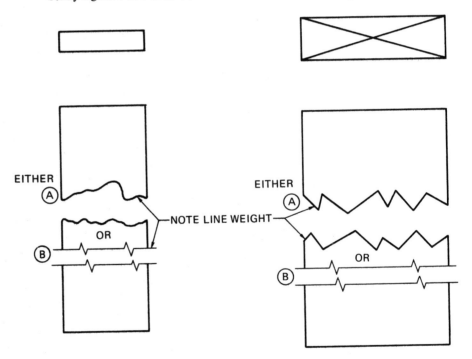

Fig. 7-19 Other types of breaks

CONVENTIONAL SECTIONING PRACTICES

Revolved Features

Clarity of a drawing can be improved by revolving a certain feature. Be sure to show the radial distances from the center for holes.

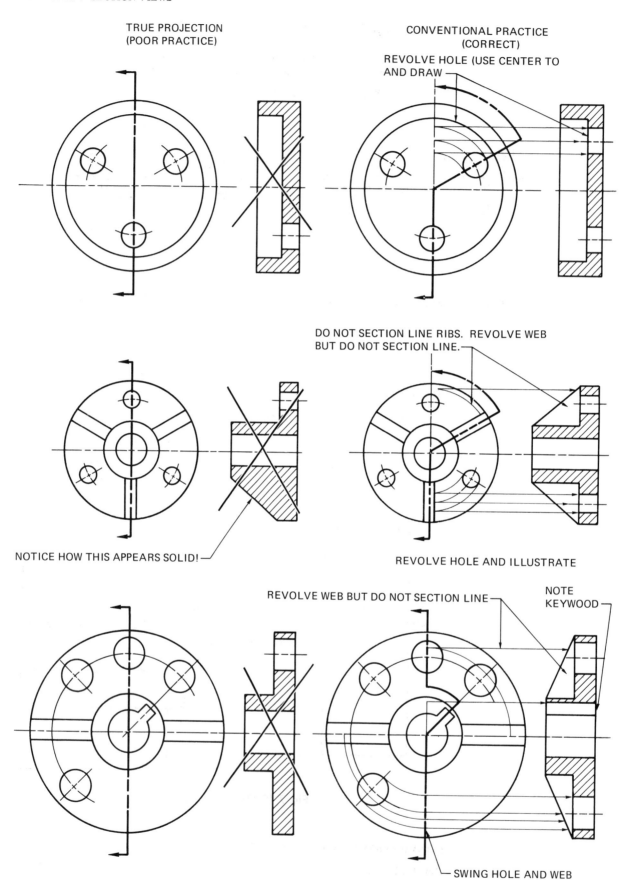

Fig. 7-20 Revolved features in section

Sectioning Ribs and Webs

True projections in section often produce incorrect impressions of the shape of the object. For example, take figure 7-21 as the object to be sectioned.

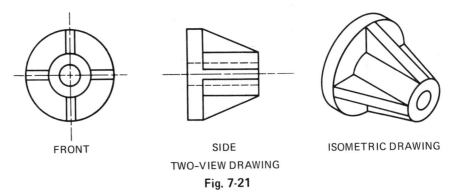

FRONT   SIDE   ISOMETRIC DRAWING

TWO-VIEW DRAWING

Fig. 7-21

A true projection of the object would look like figure 7-22.

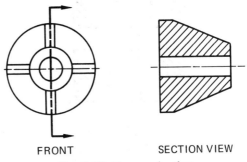

FRONT   SECTION VIEW

Fig. 7-22 True projection

This is a poor practice, however. Such a sectioned view makes the object appear to look like figure 7-23.

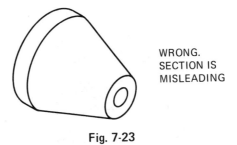

WRONG.
SECTION IS
MISLEADING

Fig. 7-23

To draw a sectioned view of figure 7-21, do not section line ribs or webs. The section view will then look like figure 7-24.

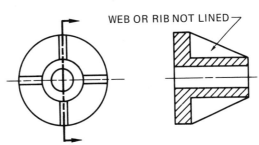

WEB OR RIB NOT LINED

Fig. 7-24

The same conventional sectioning practices apply to the example in figure 7-25.

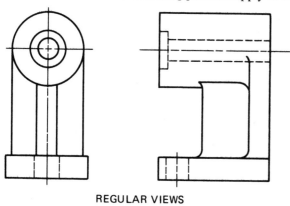

REGULAR VIEWS

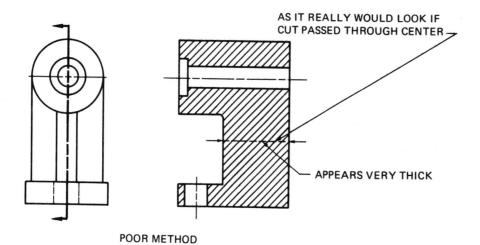

AS IT REALLY WOULD LOOK IF
CUT PASSED THROUGH CENTER

APPEARS VERY THICK

POOR METHOD

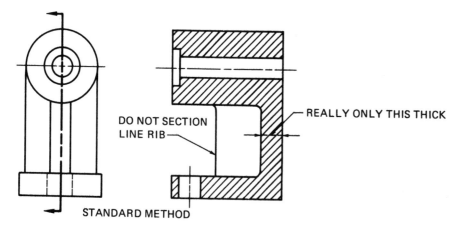

DO NOT SECTION
LINE RIB

REALLY ONLY THIS THICK

STANDARD METHOD

Fig. 7-25

CONVENTIONAL SECTIONING PRACTICES

Sometimes a drawing is made using a conventional drawing practice in order to save time and give a better description of the feature. Conventional practices are used only if the view made will be easier to read than one made from a true projection. Figures 7-26 and 7-27 show both a true projection and a conventional drawing of a pipe in section.

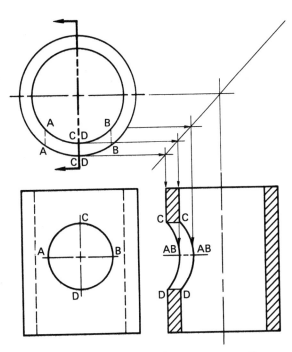

Fig. 7-26   True projection — poor practice

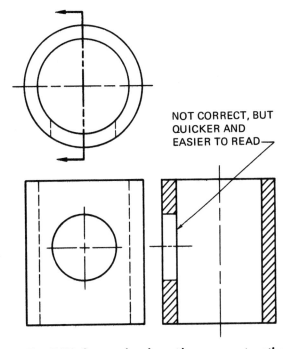

NOT CORRECT, BUT QUICKER AND EASIER TO READ

Fig. 7-27 Conventional practice — correct method

UNIT 8

DESCRIPTIVE GEOMETRY

Descriptive geometry is a strict step-by-step procedure used to layout true shape, true size, true angle(s), true distances and many other engineering functions. Chapter 9, Auxilary views, uses many of the descriptive geometry procedures to find true shape and size.

NOTATIONS

- Every view and point is labeled on a drawing by *notations*.
- All points in space are called out in lowercase letters.
  Examples: a, b, c, d, e, etc.
- Each view is called out in uppercase letters..
  Example: F = Front view   L = Left side view
             T = Top view   A = Auxiliary views
             R = Right side view B-Z = Any other views

FOLD LINES

A *fold line* indicates a 90-degree change in direction. Fold lines are illustrated by a thin, black line, figure 8-1. Figure 8-2 illustrates a regular three-view drawing without fold lines. Figure 8-3 shows a regular three-view drawing with fold lines.

- Combine uppercase letters with lowercase letters to pinpoint a point on a particular view.
  Example: "aT" = This means point "a" in the top view.
- Combine two or more points to locate a line.
  Example: Line "aT/bT" means line "a-b" in the top view.

Note when using descriptive geometry, the 45-degree projection line is not used.

Make notations at every fold line to identify the view, figure 8-4.

Line a-b is called out in each and every view, figure 8-5. Note that lower case letters are used.

FOLD LINE

Fig. 8-1

TOP

FRONT   SIDE

Fig. 8-2

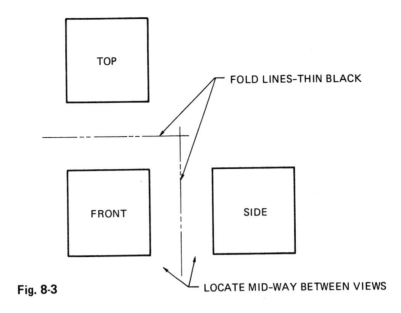

Fig. 8-3

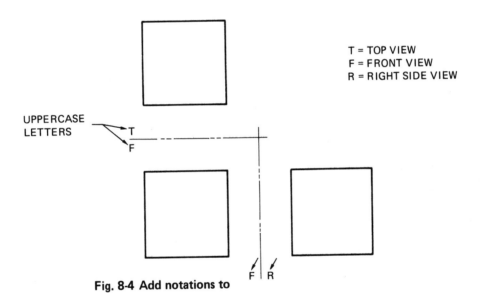

Fig. 8-4 Add notations to fold lines

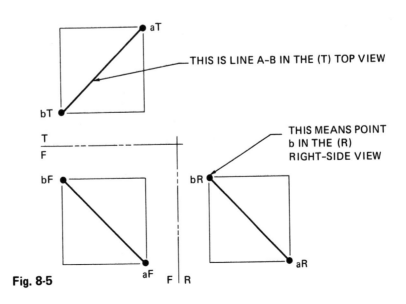

Fig. 8-5

Any point on a line must appear on that line in all views. Do not use 45-degree projection lines. Project 90 degrees from all fold lines, figure 8-6.

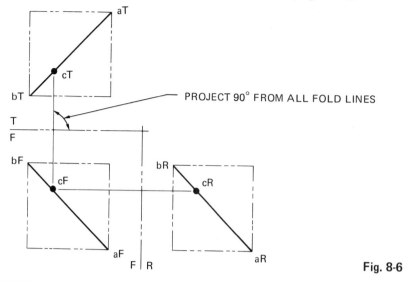

PROJECT 90° FROM ALL FOLD LINES

Fig. 8-6

PROJECTING FROM ONE VIEW TO ANOTHER

Figure 8-7 gives an example of projecting from one view to another. In this case, a right-side view is projected.

Step 1. Locate point a in the top view and measure how far it is from the fold line (dimension "x").

Step 2. Project a to the front view and across the fold line at 90 degrees, the same distance from the fold line (dimension "x").

Step 3. Do the same with point b in the top view. Project to the front and project across the same distance from the fold line. Always project at 90 degrees from the fold lines.

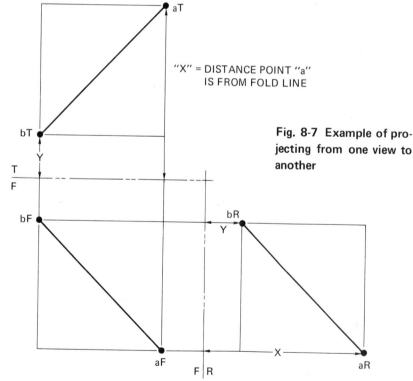

"X" = DISTANCE POINT "a" IS FROM FOLD LINE

Fig. 8-7 Example of projecting from one view to another

Rule To Remember: Always skip a view when measuring. Measure in the top view, skip the front view, transfer to the right-side view.

PROJECTING TRUE LENGTH

To find true length (T/L), an *auxiliary view* must be added. An auxiliary view is drawn from the front, top, or right view, figure 8-8.

To draw an auxiliary view:

1. Start with the regular views.
2. Add fold lines between views.
3. Label each point and each view on the fold line.
4. Draw another fold line parallel to any one of the lines in any view.
5. In the next view, it will appear as true length.
6. Remember to label all points.

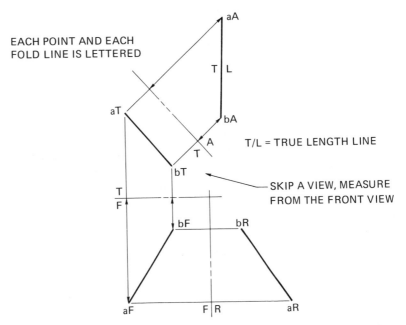

PROJECTING FROM TOP VIEW

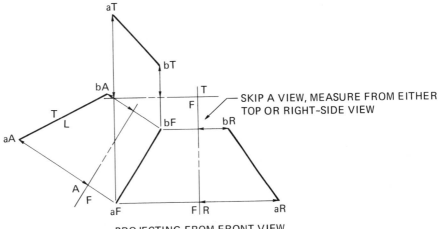

PROJECTING FROM FRONT VIEW

Fig. 8-8 Explanation for projecting true length lines

POINT VIEW

Lay a pencil on the desk and look directly down at it. You are looking at its true length. Now pick up the pencil, close one eye, and look directly at the lead end of the pencil. Notice that the pencil is no longer a line. It is now a point. This is the *point view* of the pencil. Notice also that the pencil was turned exactly 90 degrees from its true length to the point view.

Many times the point view of a line must be drawn. To draw the point view, figure 8-9, find true length, draw a fold line perpendicular (⊥) to the true length line, and draw the point view in the next view. Label each point. Skip a view to find the X length.

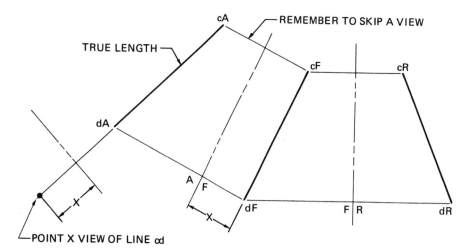

NOTE: EACH FOLD LINE EQUALS A 90° BEND.

Fig. 8-9 Finding the point view

PLANE SURFACES

A *plane surface* is a boundary which is connected by three or more points. The same steps are used to find a surface as were used to locate points in various views.

For example, given figure 8-10, draw triangle abc (△ abc) in the right-side view.

Figure 8-11 shows how to construct the right-side view. Remember to always skip a view; in this case, the front view.

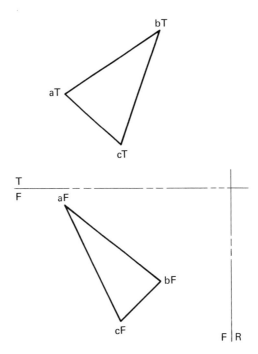

Fig. 8-10

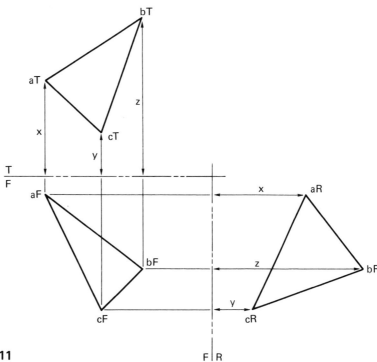

Fig. 8-11

EDGE VIEW

An *edge view* of a surface is the view one would see if the plane surface was turned on end and looked at in that position. To find an edge view, figure 8-12:

1. On any view, draw a line parallel to the fold line between views and through a point. In figure 8-12, a line is drawn in the front view, through point c, and parallel to the fold line.
2. Find the same line in the next view. It will be true length in this view.
3. Add a fold line perpendicularly to the true length line. The point view will be found in the next view as shown.
4. Add the other points to form a straight line representing the edge view.

NOTE: The surface now appears as a line and will be used later to find true shape.

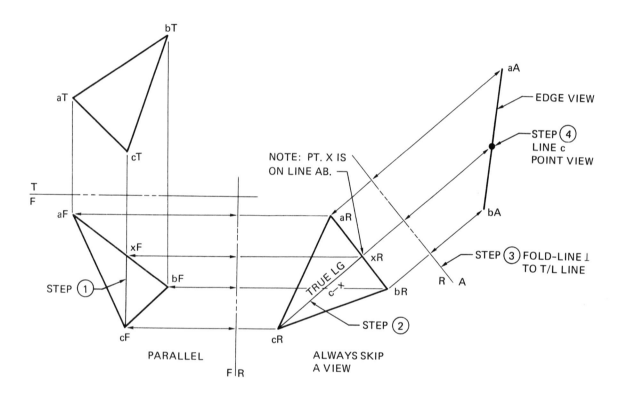

Fig. 8-12 Locating an edge view

TRUE SHAPE

The edge view must be found before drawing the *true shape*, figure 8-13. Use the same steps as those to find the edge view on page 116 and add:

5. Add fold lines parallel to the edge view.
6. In the next view the true shape will be 90 degrees from edge view. Do not forget to skip a view to get the distance.

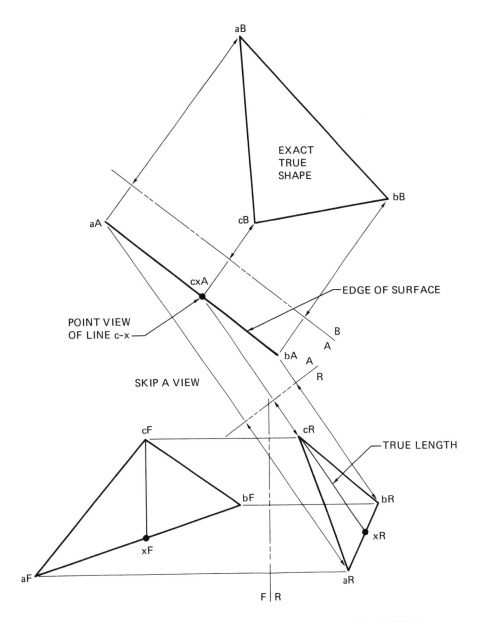

TAKE A SHEET OF PAPER AND HOLD IT UP SO ONLY THE EDGE OF IT CAN BE SEEN. ROTATE IT 90 DEGREES TO GET THE TRUE SHAPE.

Fig. 8-13 Projecting true shape

ANGLES BETWEEN TWO SURFACES

Descriptive geometry has other uses besides illustrating true shape and size. These are shown on the next few pages. Most of the same steps are used; i.e., find true length, point view, and so on. The exact angle between these surfaces must be known.

Follow these steps in order to find the angles between two surfaces, figure 8-14:

1. Add fold lines parallel to the fold.
2. Draw true length of the fold.
3. Find the point view of the fold.
4. Bring the other points along (c and d) and measure the angle between the edge view.

Always work accurately and add all notations.

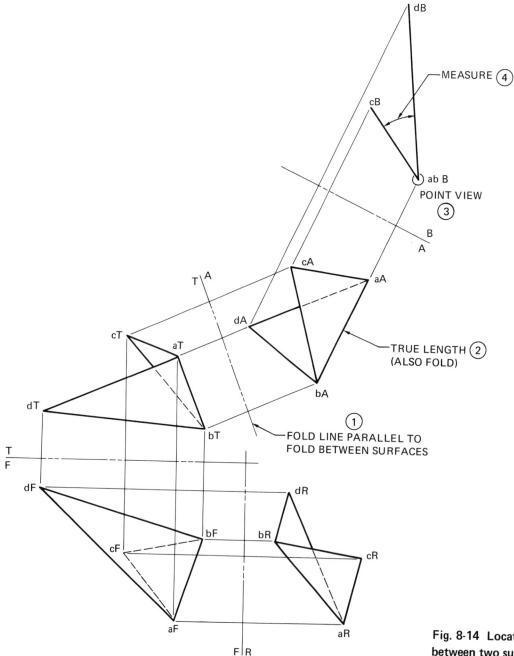

Fig. 8-14 Locating angles between two surfaces

VISIBILITY OF LINES

Visibility of lines is done by inspection and reasoning. Line aT-dT (looking into the top view, arrows 1) is closest to the fold line and appears solid in the front view. Line dT/cT is also solid. Line xT/dT is visible. Line xT/aT is between xT/bT. Therefore, in the front view, bT/xT is a hidden line. Look in one view and reason what line is closest to the fold line. In the next view it will be solid. Study figure 8-15 and apply the same reasoning to figure 8-16.

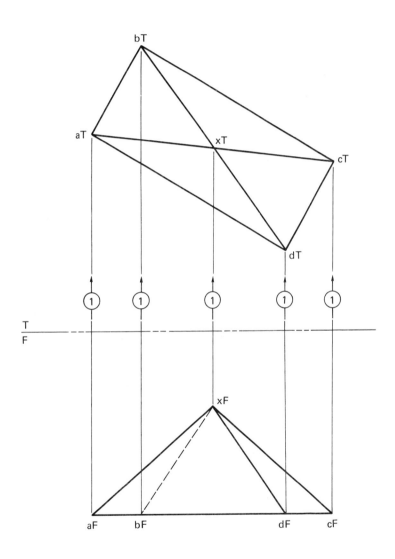

Fig. 8-15 Explanation for visibility of lines

Figure 8-16 shows two pipes: a-b and c-d. In the top view, which pipe is in front of the other? In the front view, which pipe is in front of the other?

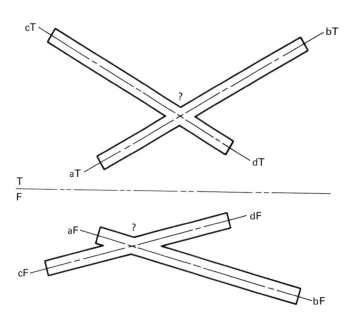

Fig. 8-16 Which pipe crosses in front of which?

Step 1. Starting from the front view where the pipes cross, draw a light line up to the top view, figure 8-17(A). Where pipe a-b crosses pipe c-d in the top view, line a-b is closer to the fold line than line c-d. Line a-b is closer to the fold line than c-d, therefore, and is a solid line in the front view.

Step 2. Use the same method for the other crossover. From the top view draw a light line down to the front view, figure 8-17(B). Where pipe a-b crosses c-d in the front view, pipe c-d is closer to the fold line than a-b. Line c-d, therefore, is in front of a-b and is a solid line in the top view.

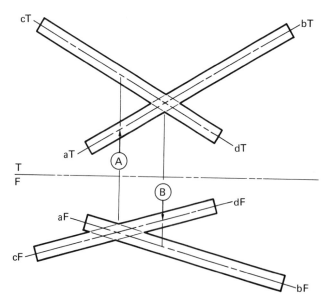

Fig. 8-17 (1) Line a-b is in front of line c-d in the front view.
(2) Line c-d is in front of line a-b in the top view.

UNIT 9

AUXILIARY VIEWS

AUXILIARY VIEWS

Many objects have inclined surfaces that are not parallel to the regular planes of projection. To show its true shape, an auxiliary view must be drawn.

An *auxiliary view* has a line of sight that is perpendicular to the inclined surface. Auxiliary views are always projected from the inclined surface at an angle of 90 degrees with the fold line. Auxiliary views give:

- True size
- True shape (or true angle)
- Points in order to draw other views

True size and shape are usually found at the same time and will be practiced first. Using auxiliary views to draw other views is illustrated at the end of this unit. Fold lines and measurements are used to lay out auxiliary views in the same manner outlined in Unit 8. Be sure to add all notations. An auxiliary view can be constructed from any view, as shown in figure 9-1.

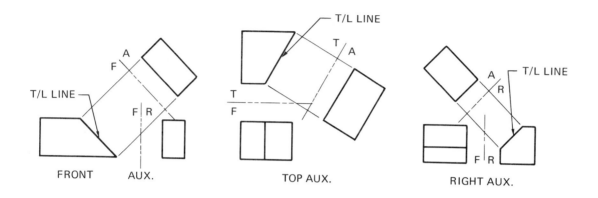

Fig. 9-1 Examples of auxiliary views

Projecting Auxiliary Views

In the front view of figure 9-2, line a/d is true length. Using descriptive geometry to find the true length line, draw a fold line parallel to the true length line. In the next view it will be the true shape and size. Do not forget to skip a view. In this case, the front view is skipped.

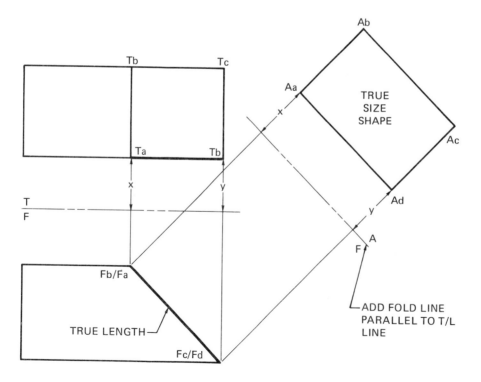

Fig. 9-2 Projecting an auxiliary view from the front view

AUXILIARY VIEW USED TO DRAW OTHER VIEWS

Auxiliary views are sometimes used to complete other views. Study the problem in figure 9-3. Follow steps one through four to complete the front view, figure 9-4.

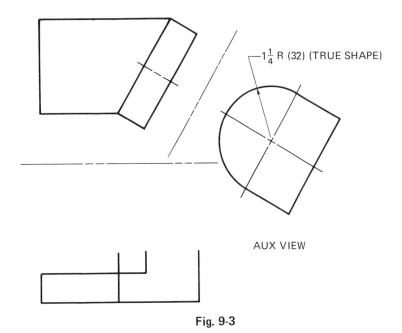

Fig. 9-3

Step 1. In the auxiliary view, divide the arc into equal segments of 30 degrees from the center.

Step 2. Project these points to the top view.

Step 3. From the top view, project these points downward.

Step 4. Using descriptive geometry methods, locate each point in the front view.

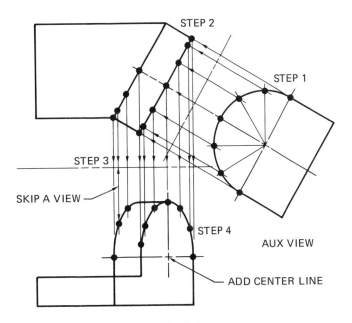

Fig. 9-4

UNIT 10

DEVELOPMENTS

DEVELOPMENTS

Development drawing is the process by which flat patterns are developed. These flat patterns are then folded or rolled to form various industrial products, such as sheet metal pipes, cardboard boxes, etc.

Many developments cannot be completed without first adding an auxiliary view to them. As in the construction of auxiliary views, developments require that the true length (T/L) of each line be determined and that each point be lettered as the work progresses. Study figure 10-1.

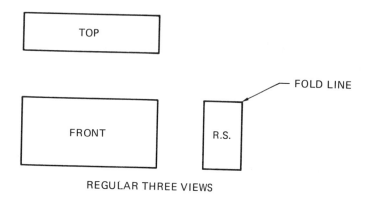

REGULAR THREE VIEWS

FOLD LINES REPRESENTED AS FOUR STRAWS LETTERED a, b, c and d

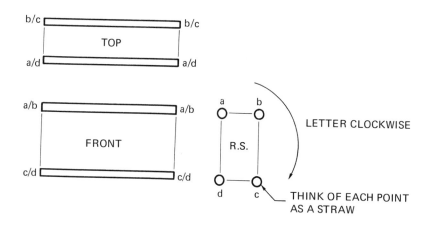

Fig. 10-1 Fold lines drawn as straws

The *base line* is any line that is 90 degrees from the straws, or fold lines. Figure 10-2 shows how to select base lines. Figure 10-3 is another example of how to draw a development. Construct the base line first with all straws, or fold lines, at 90 degrees. Neatly unroll the object, as illustrated. Note that fold line a-a is the true length, fold line b-b is the true length, etc. Each fold line is 90 degrees from the base line.

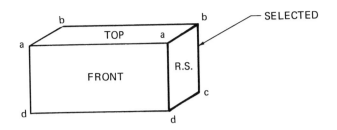

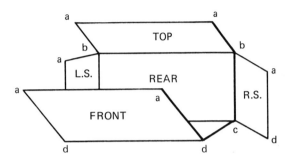

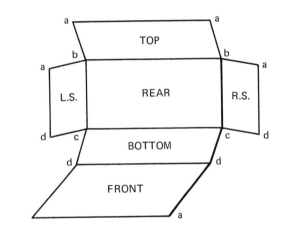

NOTE: ALL STRAWS
OR FOLD LINES
ARE PARALLEL.

NOTE: ALL FOLD LINES ARE 90°
FROM THE BASE LINE.

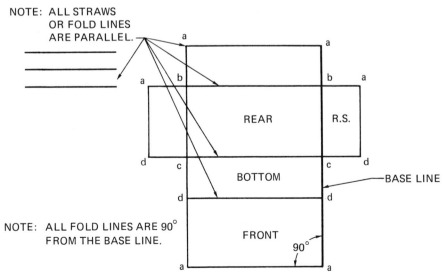

Fig. 10-2 Selecting base lines

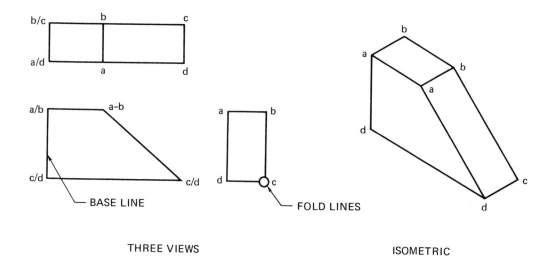

THREE VIEWS   ISOMETRIC

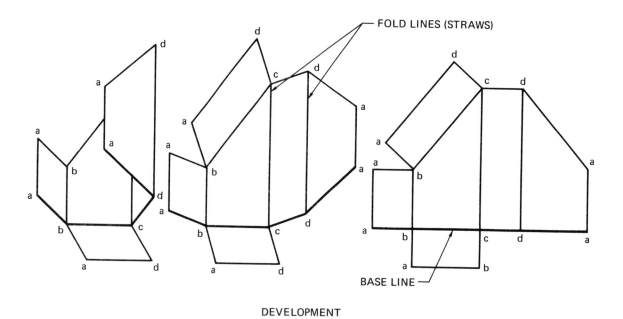

DEVELOPMENT

Fig. 10-3 Example of development

To draw a parallel line development of figure 10-4, follow steps 1 through 7.

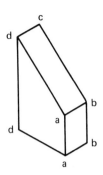

Fig. 10-4 Isometric drawing

Step 1. Draw a full-size, three-view drawing of the object. Include an auxiliary view if needed. Label each fold line or straw. Start with the shortest line.

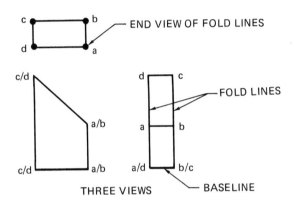

Step 2. On a separate sheet draw a base line approximately 3 inches (75) from the bottom of the paper.

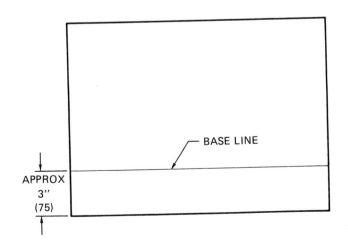

Step 3. Draw a light line approximately 1 inch (25) in from the left side of the sheet and 90 degrees from the base line. This line is the first fold line a-a. From the base line, measure up the true length of line a-a. Use the front view measurement in the three-view drawing.

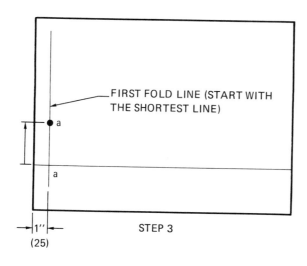

STEP 3

Step 4. From this point and along the base line, lay out the distance from a to b. Locate point b and draw a line upwards.

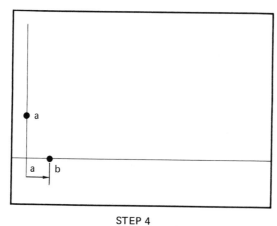

STEP 4

Step 5. Going back to the front view, measure how long line b-b is and transfer this measurment to the line starting from the base line at point b.

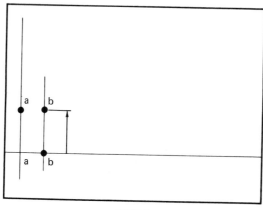

STEP 5

Step 6. Continue this process around the object and back to a-a. The order is a length, a distance, a length, a distance, etc.

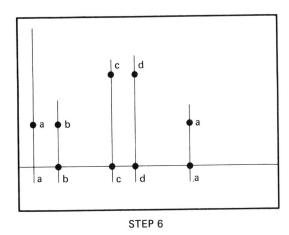

STEP 6

Step 7. Complete the development by connecting a to b, b to c, etc. Add the top and bottom (shaded area). Development is ready to be put together.

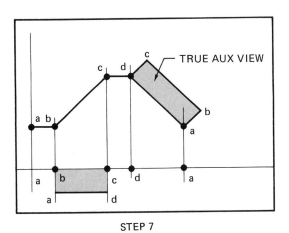

STEP 7

TABS

Tabs help hold the object together. Add tabs are from 1/16 inch (1.5) to 1/4 inch (6) wide, depending on the size of the object, and tapered 45 degrees at the ends, figure 10-5. Figure 10-6 illustrates the location of tabs. Be sure not to add tabs twice. Note tab a-d at the bottom, X . Do not add another tab at a-a, Y .

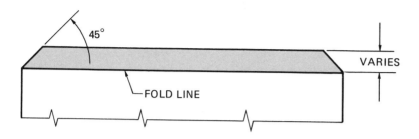

Fig. 10-5 Typical tab

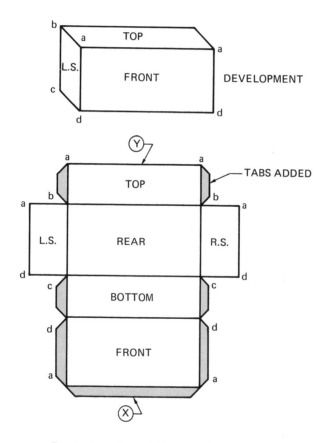

Fig. 10-6 Tabs added to development layout

DEVELOPMENTS OF ROUND OBJECTS

Round or partly round objects are drawn the same way as parallel line developments except that the drafter must choose the fold line locations. Number each point as illustrated in figure 10-7.

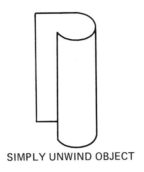

SIMPLY UNWIND OBJECT

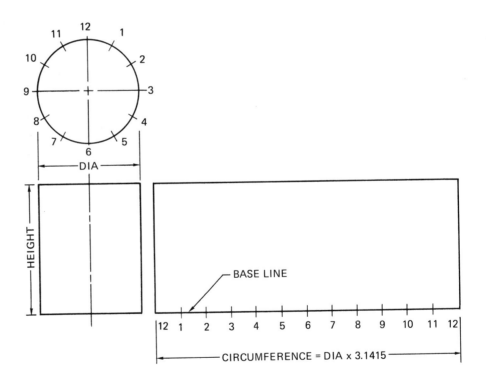

NOTE: SEE APPENDIX B FOR THE CIRCUMFERENCE AND AREA OF CIRCLES

Fig. 10-7 Development of round object

Compare figure 10-8 with figure 10-1. They are the same except for the addition of fold lines to the curved area. Proceed with the layout using the same method of development. Add fold lines to curved surface equally spaced at 30 degrees.

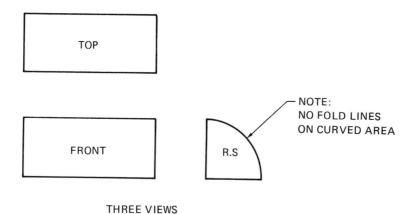

THREE VIEWS

ADD FOLD LINES TO CURVED SURFACE EQUALLY SPACED AT 30 DEGREES

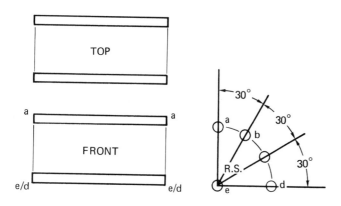

ADD FOLD LINES AT EQUAL SPACES, USUALLY
10°, 20°, OR 30°. LETTER CLOCKWISE.

Fig. 10-8 Development of partly round object

CURVED AND NOTCHED TABS

When a surface is round or curved, the tabs must be notched, figure 10-9.

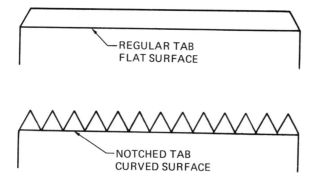

Fig. 10-9 Regular and notched tabs

RADIAL LINE DEVELOPMENT

The same basic steps are used to construct a radial line development as those used in a parallel line development. The main difference is that all fold lines on a *radial line development* go to one point. They are not parallel as in parallel line developments.

A pyramid is a good example of an object that is made by the radial line method. In figure 10-10 all four sides go to point X.

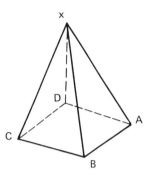

Fig. 10-10 Pyramid

Remember that with radial line development, the true length of each line must be found. The drafter cannot determine by looking at figure 10-11 what the true length of line AX is in either the top or the front view.

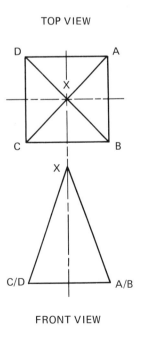

Fig. 10-11 Two views of the pyramid

Revolve the object, figure 10-12, until line AC is on center line A'C'. At this position X-A' is a true length. To determine the true length in the front view, extend the base line to the left until it meets a line dropped from A' in the top view. Where these lines meet locates A' in the front view. Measure distance X-A' in the front view to determine the true length needed for the development.

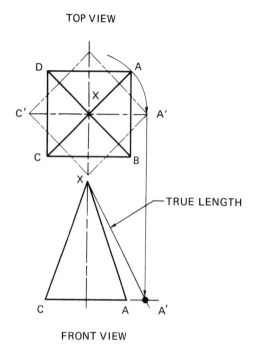

Fig. 10-12 Finding the true length of a pyramid's side

Development of a Pyramid

To make a development of the pyramid, follow steps 1 through 6, figure 10-13.

Step 1. Find the true length of the object, line X-A'.

Step 2. Swing an arc with the true length as radius.

Step 3. Place point A anywhere on that arc.

Step 4. Draw a line from A to X.

Step 5. Step off distances AB, BC, CD, and CA. Connect these points with point X.

Step 6. Add the bottom. To do this, draw perpendiculars to line BC at points B and C. Swing an arc from point D using CD as the radius. Construct a parallel line to BC running through point D.

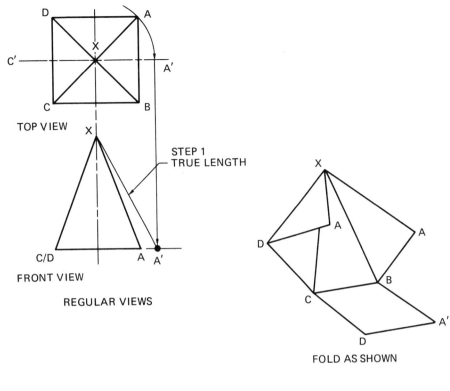

TOP VIEW

FRONT VIEW

REGULAR VIEWS

STEP 1
TRUE LENGTH

FOLD AS SHOWN

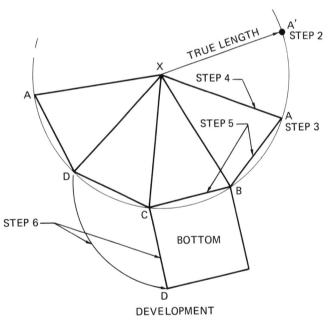

TRUE LENGTH

STEP 2

STEP 4

STEP 5

STEP 3

STEP 6

BOTTOM

DEVELOPMENT

Fig. 10-13 Development of a pyramid using the radial-line method

Development of a Cone

Steps 1 through 6 outline how to make a development of a cone, figure 10-14.

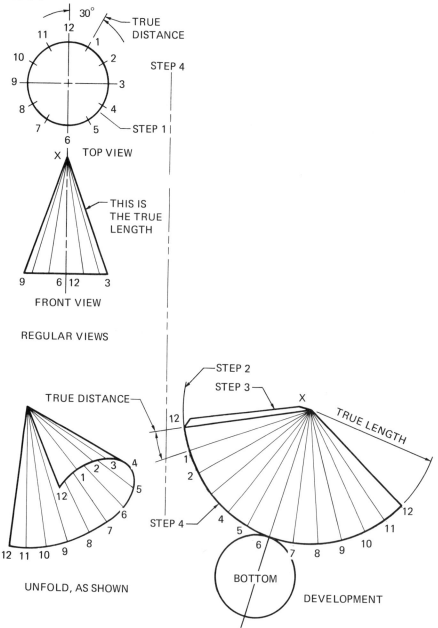

Fig. 10-14 Development of a cone

Step 1. Divide the given circle into 30-degree segments. Number each point.

Step 2. With a radius equal to the true length (line X-3 in the front view), swing an arc.

Step 3. Place point 12 anywhere on that arc. Connect points 12 and X.

Step 4. Measure the true distance between points 12-1 in the top view. Step this distance off on the arc from point 12.

Step 5. Along the arc, step off all 12 parts of the circle. Connect the last point, a new point 12, with point X. This completes the body development of the cone.

Step 6. To add the bottom, make a tangent circle to any point on the cone's bottom arc equal to the diameter of the given circle found in the top view.

MEASURING TRUE LENGTHS IN A DEVELOPMENT

Study figure 10-15. Only lines 3 and 9 are the true length; the others are not. Looking at the front view, each line starts from point x at the top and comes downward. Thus, they cannot be true length. In order to find true length, rotate the drawing six times, figure 10-15, and project each line to one side or the other, figure 10-16. Note how each point is projected to the left. This is exactly the same as drawing each rotated point. Now that the true length of each line is known, make a normal radial line development. Number each point.

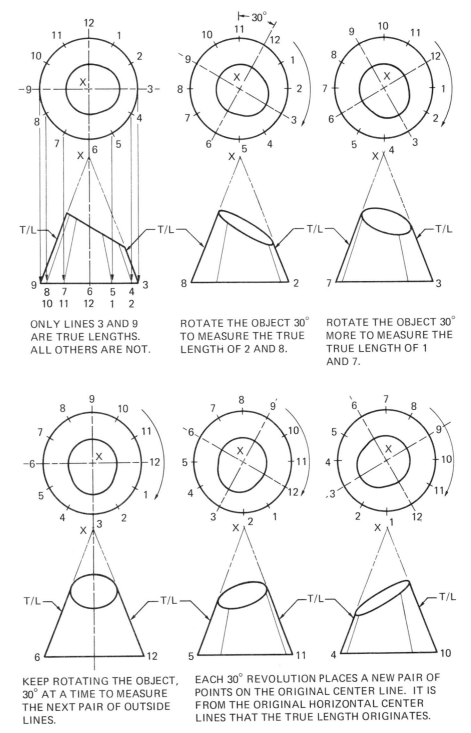

ONLY LINES 3 AND 9 ARE TRUE LENGTHS. ALL OTHERS ARE NOT.

ROTATE THE OBJECT 30° TO MEASURE THE TRUE LENGTH OF 2 AND 8.

ROTATE THE OBJECT 30° MORE TO MEASURE THE TRUE LENGTH OF 1 AND 7.

KEEP ROTATING THE OBJECT, 30° AT A TIME TO MEASURE THE NEXT PAIR OF OUTSIDE LINES.

EACH 30° REVOLUTION PLACES A NEW PAIR OF POINTS ON THE ORIGINAL CENTER LINE. IT IS FROM THE ORIGINAL HORIZONTAL CENTER LINES THAT THE TRUE LENGTH ORIGINATES.

Fig. 10-15

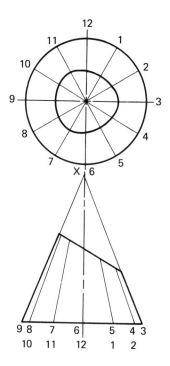

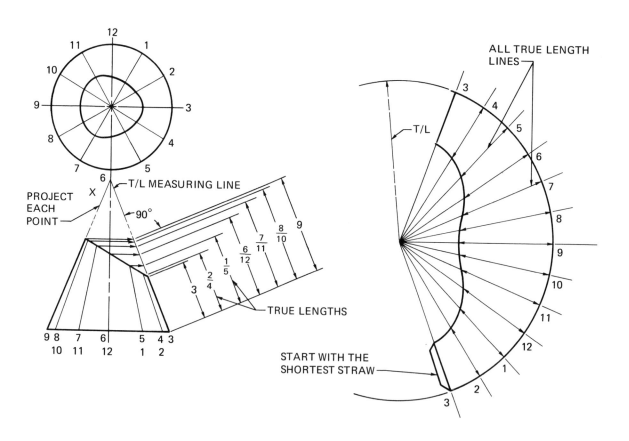

Fig. 10-16

TRIANGULATION DEVELOPMENT

A *triangulation development* is done by breaking the surface into a series of triangles. Each example in figure 10-17 is developed using triangulation. Note the triangular shapes in the objects. Each triangle must be drawn.

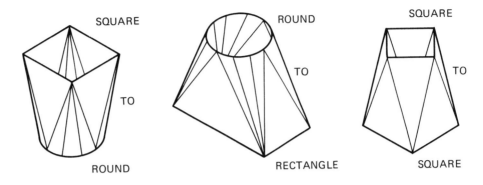

Fig. 10-17 Typical objects that are developed using triangulation

To draw each triangle, the true length of each leg is found. The triangles are connected together to form the development. Be sure to use true lengths. To find true lengths, a true length diagram must be made.

True Length Diagram

The true length diagram is a quick way to find the true length of all lines. Descriptive geometry may be used, but this method usually is faster. True length is a combination of two views, top and front, figure 10-18. Figure 10-19 outlines the method for making a true length diagram.

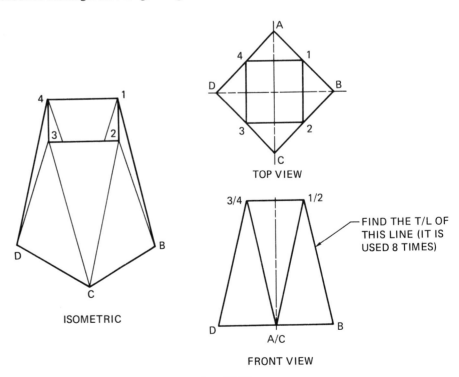

Fig. 10-18

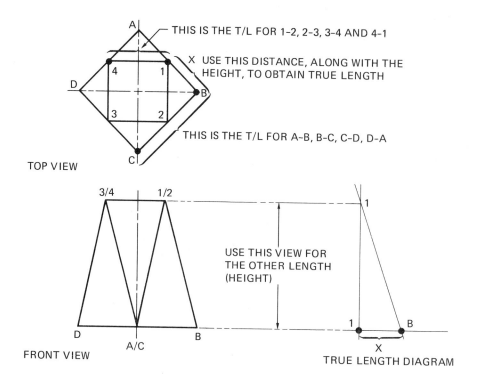

TOP VIEW

THIS IS THE T/L FOR 1-2, 2-3, 3-4 AND 4-1

X USE THIS DISTANCE, ALONG WITH THE HEIGHT, TO OBTAIN TRUE LENGTH

THIS IS THE T/L FOR A-B, B-C, C-D, D-A

FRONT VIEW

USE THIS VIEW FOR THE OTHER LENGTH (HEIGHT)

TRUE LENGTH DIAGRAM

Fig. 10-19 True length diagram

Once all the true lengths are found, draw triangle 1-B-2, triangle B-2-C, triangle 2-C-3, and so on all around to line A-1, figure 10-20. Label each point.

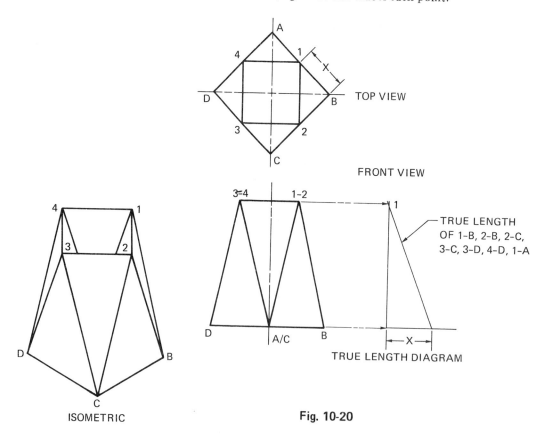

TOP VIEW

FRONT VIEW

TRUE LENGTH OF 1-B, 2-B, 2-C, 3-C, 3-D, 4-D, 1-A

TRUE LENGTH DIAGRAM

ISOMETRIC

Fig. 10-20

Using the true lengths of all lines, start with line 1-B and construct triangle 1-B-2, figure 10-21 (steps 1, 2, and 3). From point B, swing arc BC. From point 2, swing arc 2-C. Where the two arcs cross is point C. Work around the development in this manner. Use the true lengths each step of the way. Label each point as it is found. Visualize the steps. Draw an object following the steps as illustrated. Do not proceed until the sketch is understood.

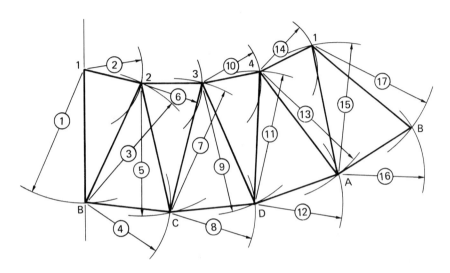

Fig. 10-21

DEVELOPMENT OF AN OBJECT WITHOUT FOLDS

Sometimes an object does not have folds. One shape just flows into another shape. In figure 10-22, the top of the object is round and flows into an ellipse at the bottom.

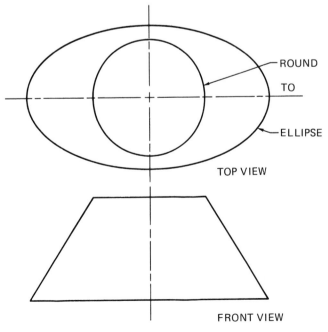

ROUND
TO
ELLIPSE

TOP VIEW

FRONT VIEW

Fig. 10-22

To develop this object, triangles must be added before proceeding as if it was a normal triangulation problem.   Steps 1 through 7 outline the method, figure 10-23.

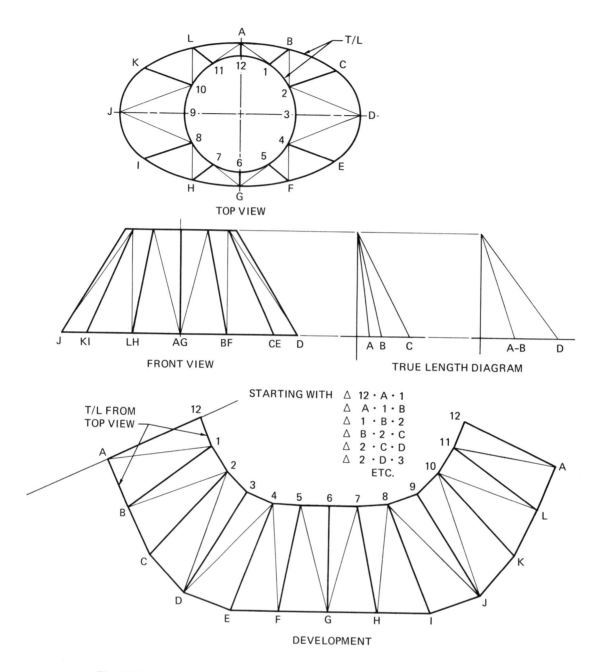

Fig. 10-23 Triangulation development, round to ellipse

Step 1. Divide the top into even spaces.
Step 2. Divide the bottom into the same amount of spaces.
Step 3. Number and letter each point.
Step 4. Draw the solid lines.
Step 5. Draw the dashed lines.
Step 6. Make two true length diagrams.
Step 7. Lay out development using all true lengths.

BEND ALLOWANCE

A drafter must determine the true length of an object before it is made or bent. This is called the *developed length*. In order to determine this length, the distance needed to make the bend is found. Charts are available for this purpose. The chart in appendix C-1 (appendix C-2 for metric), for example, is used if the bend is at 90 degrees. The chart in appendix C-3 (appendix C-4 for metric) is used if the bend is at an angle other than 90 degrees.

To determine the length of the object in figure 10-24:

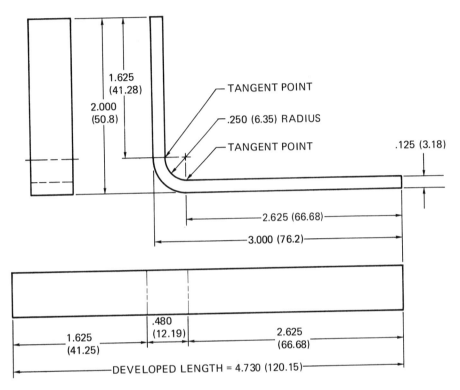

Note: Tangent means to or where the straight section stops and the curved section starts (or stops)

Fig. 10-24 Determining developed length for an object bent at 90 degrees. For metric figures use appendix C-2.

Step 1. Locate the tangent points at the ends of the .250-inch (6.35) radius (inside radius).

Step 2. Determine the thickness of the stock from the drawing: .125 inch (3.18)

Step 3. Refer to appendix C-1. The left-hand column gives the thicknesses while radii are given across the top. Follow the column down to .125 (3.18) and to the right to the column with the heading of .250 (6.35). Where these two columns cross is the distance that must be added to get a 90-degree bend: .480 inch (12.20).

Step 4. Add this measurement to the distances from the tangent points to the end of the stock and the developed length will be determined:

inches: 1.625 + .480 + 2.625 = 4.730
millimetres: 41.28 + 12.19 + 66.68 = 120.15

If the object is more or less than 90 degrees, use appendix C-3 (appendix C-4 for metric), the same way as the previous chart. Multiply the number found on the chart by the distance around the bend in degrees. In figure 10-25, 60° + 90° = 150°. Then: 150° × .00534 = 0.801. Add this to the straight areas. The answer is 4.951 developed length.

IMPORTANT! DO NOT USE
THE INSIDE DEGREE FIG.
FIGURE OUT HOW FAR
AROUND THE BEND GOES
90° + 60° = 150°

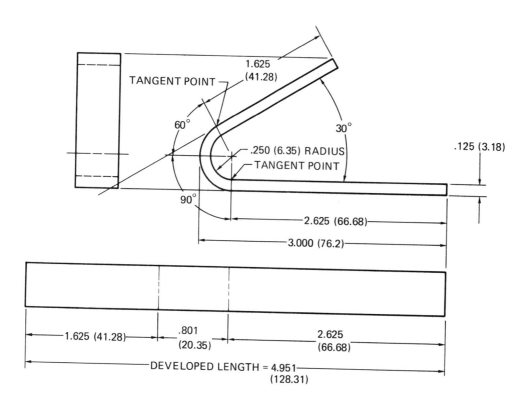

Fig. 10-25 Determining developed length for an object bent at other than 90 degrees. For metric figures use appendix C-4.

Determining Developed Length

Many required dimensions are missing from the sketch in figure 10-26. Do not forget that the tangent point is where the straight area stops and the radius starts.

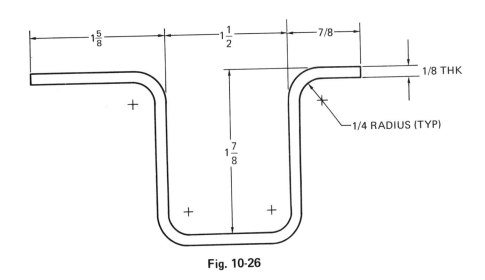

Fig. 10-26

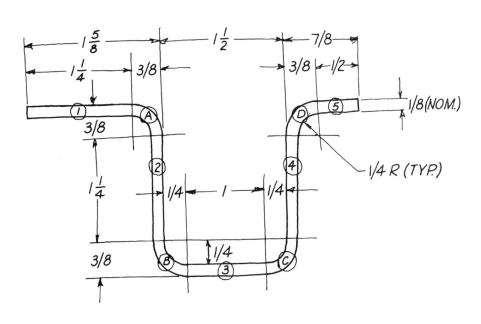

Fig. 10-27 Determining developed length

To determine the developed length, figure 10-27:

Step 1. Make a neat sketch. Try to keep it in scale.
Step 2. Add all dimensions to the sketch, even ones not needed.
Step 3. From one end, number each straight area (1 through 5).

Step 4. From the same end, letter every bend (A-D).

Step 5. Add all the straight lengths together:

$$
\begin{aligned}
1 &= 1\ 1/4 \\
2 &= 1\ 1/4 \\
3 &= 1 \\
4 &= 1\ 1/4 \\
5 &= \underline{\ \ 1/2} \\
&\ 5\ 1/4\ \text{total of straight lengths}
\end{aligned}
$$

Step 6. Add all the bends together:

$$
\begin{aligned}
A &= .480 \\
B &= .480 \\
C &= .480 \qquad \text{(See Appendix C-1, 90}^\circ\text{ bends)} \\
D &= \underline{.480} \\
&\ 1.920\ \text{total all bends}
\end{aligned}
$$

Step 7. Add the totals. Label each step.

$$
\begin{aligned}
&5.250\ \text{All straight lengths} \\
&\underline{1.920}\ \text{All bend allowances} \\
&7.170\ \text{Total developed length}
\end{aligned}
$$

KINDS OF SEAMS

Refer to appendix D for a list of the standard thicknesses of sheet metal.

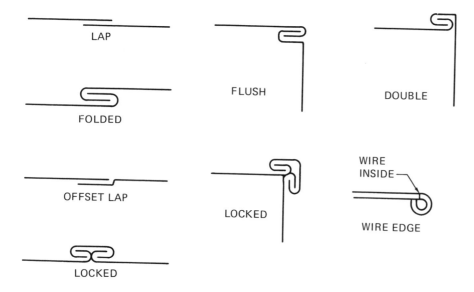

Fig. 10-28   Types of sheet metal seams

UNIT 11

BASIC DIMENSIONING

DIMENSIONING

Previous units have covered correct line weight, neatness, accuracy, and general constructions. To be of any value, however, it is most important that a drawing be dimensioned properly. This unit covers the rules of standard dimensioning.

Drafters should place themselves in the position of the person reading the drawing and dimension the drawing accordingly. If a drafter is in doubt whether a dimension is needed, it should be included. Other basic practices are:

- It should not be necessary to scale a drawing to determine a dimension.

- It should not be necessary to calculate to determine a dimension.

- It should not be necessary to assume anything to determine a dimension.

DIMENSIONING SYSTEMS

The two systems used to dimension a drawing are: aligned and unidirectional. In an *aligned system* all dimensions are read from the bottom and right side of the page, figure 11-1. This is the old system. All dimensions are in line (aligned) with the dimension line.

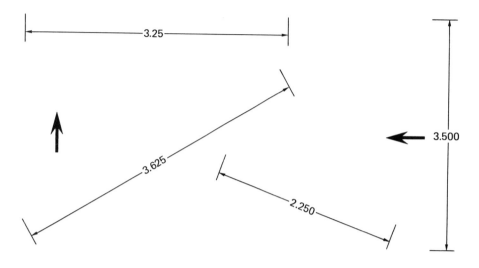

Fig. 11-1 Aligned dimensioning

In a *unidirectional system* all dimensions are read from the bottom of the page, figure 11-2. This is the system currently preferred.

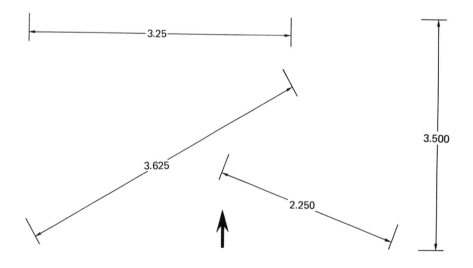

Fig. 11-2 Unidirectional dimensioning

ARROWHEADS

A good arrowhead proportion is about 1/8 inch (3) long and has a width equal to approximately 1/3 of the length. Practice making arrowheads using figures 11-3 and 11-4 as a guide.

Fig. 11-3 Arrowhead proportions   Fig. 11-4 Steps in making an arrowhead

RULES FOR DIMENSIONING

The general rule when dimensioning is to place the first dimension a minimum of 5/8 inch (15) away from the object, figure 12-5. All succeeding dimensions are spaced a minimum of 3/8 inch (9) apart. There is a visible gap between extension lines and the object that extends approximately 1/8 inch (3) past the last dimension line. This is important as it will enhance the quality of the drawing and its comprehension.

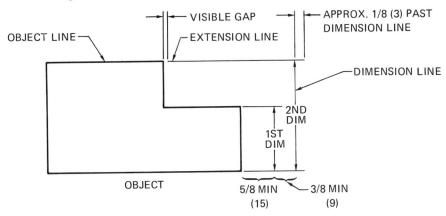

Fig. 11-5 Placement of dimensions

Dimension Lines

The general rule is to place the shortest dimension line closest to the object, figure 11-6 at right, so extension lines do not cross any more than necessary.

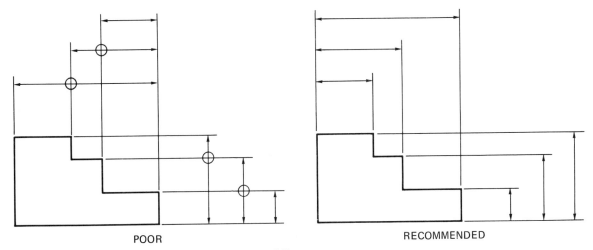

POOR   RECOMMENDED

Fig. 11-6 Placement of lines

Leaders

The *leader* is an inclined straight line with a short, horizontal portion extending to the midpoint of the note or dimension, figure 11-7. They are drawn at an angle of 30 to 60 degrees. On a circle, the leader starts at the edge of the circle on a line that projects to the center.

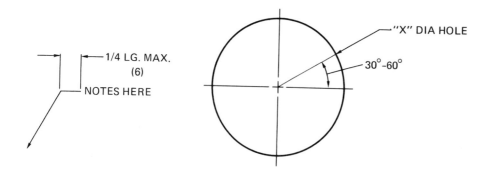

Fig. 11-7 Placement of leaders

Basic Shapes

When dimensioning, think about the size of the basic shape and how it will be machined. Most drawings have width (WD), height (HGT), and depth (DP). Start with these and try to place them between views, figure 11-8.

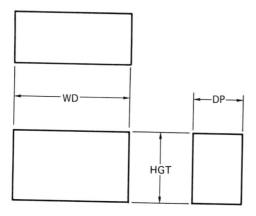

Fig. 11-8

Arrowheads

A general rule for arrowheads when dimensioning radii and diameters:

- Radii — arrowheads inside
- Diameters — arrowheads outside

Figure 11-9 shows how the arrowheads are placed on a drawing. Do not dimension to a hidden line or to the center line of a hidden circle.

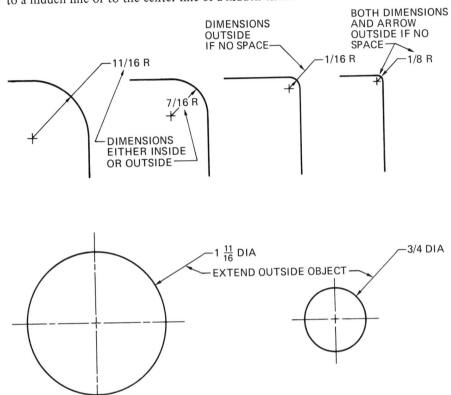

Fig. 11-9 Placement of arrowheads

Unnecessary Dimensions

Always omit unnecessary dimensions. The missing dimension in figure 11-10 does not need to be there. If it is shown it must be identified as a REF dimension.

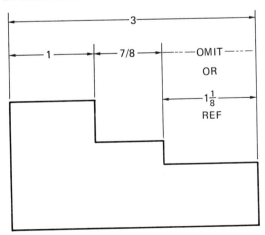

Fig. 11-10 Proper dimensioning

Changing a Dimension

Most changes made on a drawing are not major enough to warrant redoing the entire drawing. If just a dimension or two needs to be changed, correct the dimension(s) to the size required and place a wavy line under the corrected dimension. This indicates the dimension is correct but out-of-scale, figure 11-11.

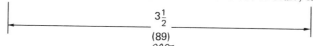

$$3\frac{1}{2}$$
(89)

Fig. 11-11 Correcting a dimension on a drawing

Round Objects

Good drafters have an idea how things are manufactured and a basic knowledge of how simple tools are used. They must keep in mind how the object will be made. If the object is round, it probably will be put into a lathe and turned down. The machinist must know the diameter, therefore, not the radius, figure 11-12.

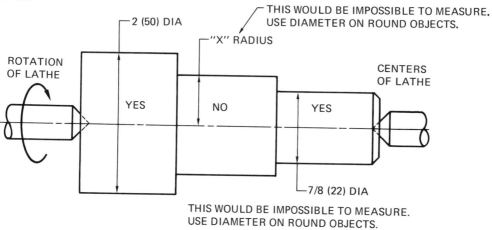

THIS WOULD BE IMPOSSIBLE TO MEASURE. USE DIAMETER ON ROUND OBJECTS.

2 (50) DIA

"X" RADIUS

ROTATION OF LATHE

CENTERS OF LATHE

YES   NO   YES

7/8 (22) DIA

THIS WOULD BE IMPOSSIBLE TO MEASURE. USE DIAMETER ON ROUND OBJECTS.

Fig. 11-12 Dimensioning round objects

Holes

Holes in an object are called out by diameters, not by radii. It is important that the center lines of holes be located accurately. The machinist must know exactly where to start the center of the hole, figure 11-13.

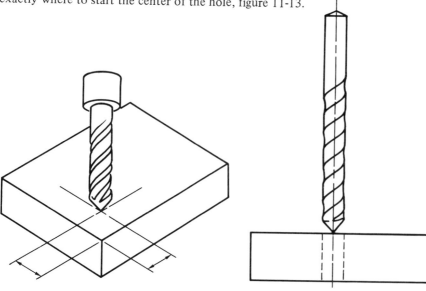

Fig. 11-13 Dimensioning hole center

Reference Dimensions

Reference dimensions are marked on a drawing with the symbol REF. They are given for information only and are not used in the manufacturing process, figure 11-14. When used, the sum of incremental dimensions must equal the overall dimension.

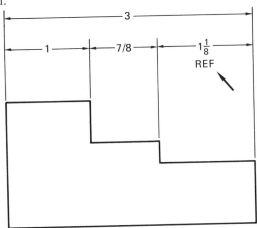

Fig. 11-14 Reference dimensioning

Chamfers

A *chamfer* is a beveled edge. When a chamfer is cut at a 45-degree angle, it is dimensioned as in figure 11-15, diagram A. If the chamfer is not cut at a 45-degree angle, it is dimensioned as in diagram B.

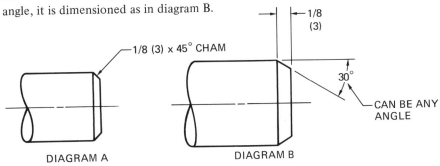

Fig. 11-15 Chamfer dimensioning

Knurling

The process of rolling depressions of either straight or diamond designs on cylindrical surfaces is called *knurling*. The finer the knurl, the more grooves per inch it will have. Be sure to indicate the kind of knurl and its pitch. Knurls are used for appearance and functional purposes. Figure 11-16 shows what knurled surfaces look like.

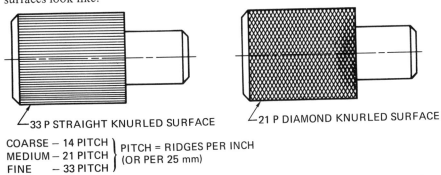

Fig. 11-16 Dimensioning knurled surfaces

SIZE AND LOCATION

The two kinds of dimensions are *size (S) dimensions,* which describe the size of a feature, and *location (L) dimensions,* which specify the location of a feature, figures 11-17 and 11-18.

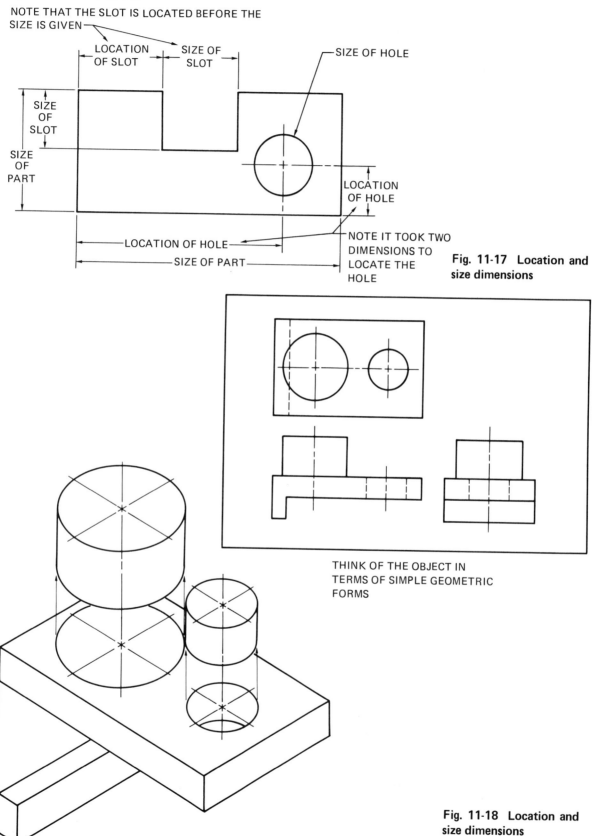

NOTE THAT THE SLOT IS LOCATED BEFORE THE SIZE IS GIVEN

LOCATION OF SLOT

SIZE OF SLOT

SIZE OF HOLE

SIZE OF SLOT

SIZE OF PART

LOCATION OF HOLE

LOCATION OF HOLE

SIZE OF PART

NOTE IT TOOK TWO DIMENSIONS TO LOCATE THE HOLE

Fig. 11-17   Location and size dimensions

THINK OF THE OBJECT IN TERMS OF SIMPLE GEOMETRIC FORMS

Fig. 11-18   Location and size dimensions

Callouts for Holes

A drafter must know and understand how various types of holes are made and how to call out each type on a drawing. The five most commonly used methods are the *drill, ream, counterbore, countersink,* and *spot face.* Each is illustrated in figure 11-19 with the correct callout. Note that each hole is drilled first. A reamed hole, for instance, must be drilled undersize before reaming. Do not name the process used as that is up to the worker. Just call out the tolerance required.

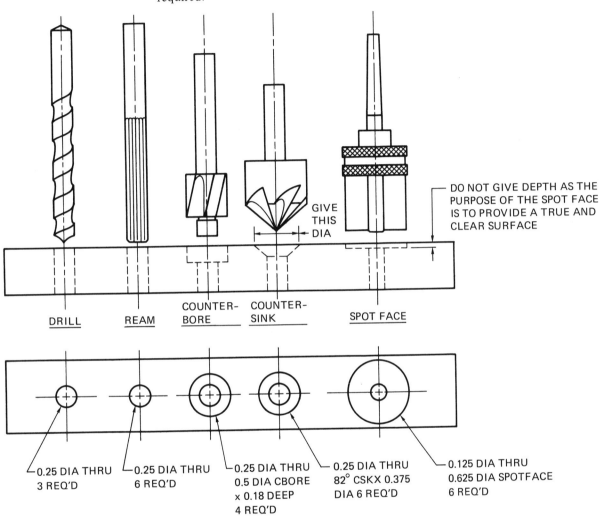

DRILL   REAM   COUNTER-BORE   COUNTER-SINK   SPOT FACE

GIVE THIS DIA

DO NOT GIVE DEPTH AS THE PURPOSE OF THE SPOT FACE IS TO PROVIDE A TRUE AND CLEAR SURFACE

0.25 DIA THRU 3 REQ'D

0.25 DIA THRU 6 REQ'D

0.25 DIA THRU 0.5 DIA CBORE x 0.18 DEEP 4 REQ'D

0.25 DIA THRU 82° CSK X 0.375 DIA 6 REQ'D

0.125 DIA THRU 0.625 DIA SPOTFACE 6 REQ'D

Fig. 11-19 Order of callout for dimensioning holes

Bolts and Nuts

Figure 11-20 shows a bolt and nut.

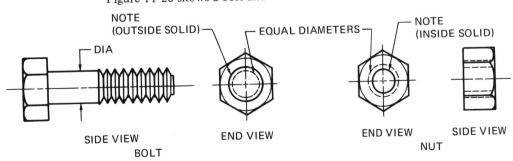

NOTE (OUTSIDE SOLID)   EQUAL DIAMETERS   NOTE (INSIDE SOLID)

DIA

SIDE VIEW   END VIEW   END VIEW   SIDE VIEW
BOLT   NUT

Fig. 11-20 Dimensioning bolts and nuts

Note the diameter of the bolt in figure 11-21. The thread or root is smaller than the outside diameter. When an external thread is drawn on a bolt or rod, it is drawn as in figure 11-21. Note the nut in figure 11-22. The diameter of the bolt is now a hidden line and the inside of the threads are now a solid line.

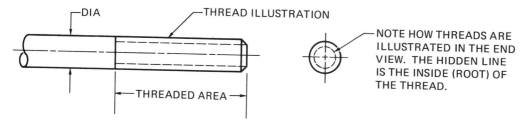

Fig. 11-21 External threads

Fig. 11-22 Internal threads

SPECIFYING MACHINED SURFACES

A *finished surface* is a surface that has been machined for functional or appearance purposes. *Finish texture symbol* shows that a surface is to be finished. The symbol has changed over the years to specify closer tolerances, figure 11-23.

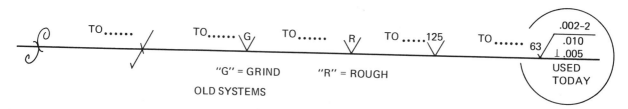

Fig. 11-23 Evolution of finish symbols

A finish texture symbol or mark is made the size and shape indicated in figure 11-24.

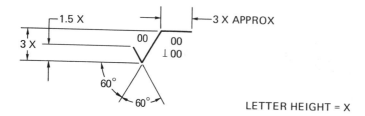

Fig. 11-24

It may be drawn freehand or with a template. A number equaling the roughness required in microinches or micrometres is neatly lettered within the angle of dimension. One microinch equals .000001 of an inch. One micrometre equals 0.000001 metre, figures 11-25 and 11-26.

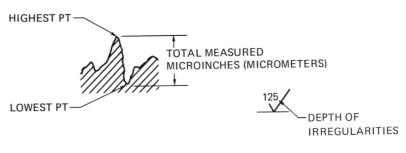

$$\begin{array}{r} .000001 \\ \times \quad 125 \\ \hline .000125 \end{array}$$ DEPTH OF
SURFACE
IRREGULARITY

Fig. 11-25 Surface irregularities

ROUGHNESS		KIND OF SURFACE	USAGE
μm	μin.		
12.5	500	Rough	Used where vibration or stress concentration are not critical and close tolerances are not required.
6.3	250	Medium	For general use where stress requirements and appearance are of minimal importance.
3.2	125	Average smooth	For mating surfaces of parts held together by bolts and rivets with no motion between them.
1.6	63	Smoother than average finish	For close fits or stressed parts except rotating shafts, axles, and parts subject to vibrations.
0.8	32	Fine finish	Used for such applications as bearings.
0.4	16	Very fine finish	Used where smoothness is of primary importance such as high-speed shaft bearings.
0.2	8	Extremely fine finish	Use for such parts as surfaces of cylinders (engines).
0.1	4	Super fine finish	Used on areas where surfaces slide and lubrication is not dependable.

Fig. 11-26 Roughness chart for common finished surface μm = micrometre
μin. = microinch

Finished Surface Rules

1. Edge view surfaces appearing in more than one view have finish marks on each view.
2. FAO is the abbreviation for "finish all over."
3. Know the types of surfaces and their general usages. 125 (3.2) is average.
4. Do not place finish marks on a hole surface unless a special finish is required, such as a reamed hole.

Dimensions should originate from an important surface, such as a mating surface (usually a finished surface), an important center line, or an important point feature on an object. This is called *base line dimensioning*. Study figure 11-27. The finished surface at left is the starting surface for all dimensions going from left to right. The finished surface at the bottom is the starting surface for all vertical dimensions. Locate a feature, then indicate its size.

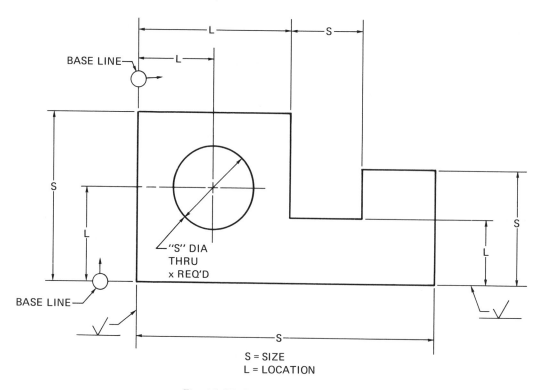

Fig. 11-27 Surface dimensioning

Bolt Circles

Study figure 11-28. A and B show how to dimension equally spaced holes on the same bolt circle (BC). C shows how to dimension using angles about a common center. D shows how to dimension using coordinates about a common center. E shows how to locate holes on a common center by polar coordinates. The method used depends on the accuracy intended to be achieved.

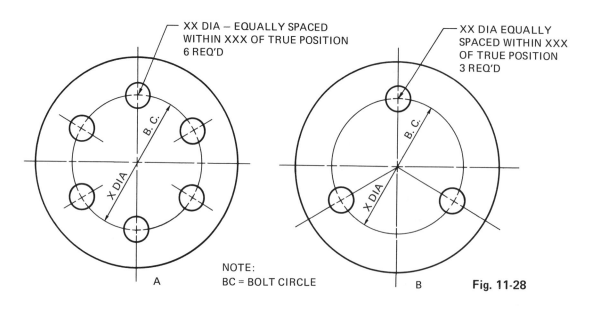

NOTE:
BC = BOLT CIRCLE

Fig. 11-28

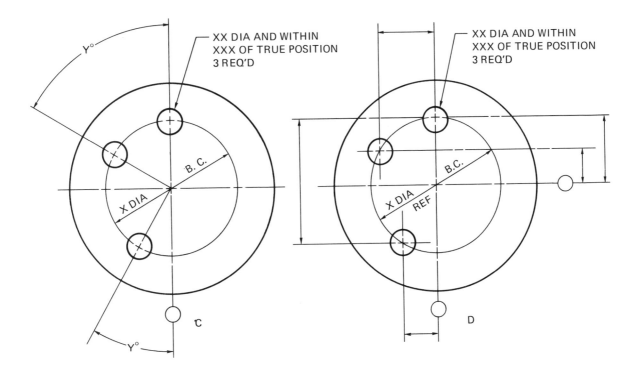

XX DIA AND WITHIN
XXX OF TRUE POSITION
3 REQ'D

B.C.

X DIA

Y°

Y°

C

XX DIA AND WITHIN
XXX OF TRUE POSITION
3 REQ'D

B.C.

X DIA REF

D

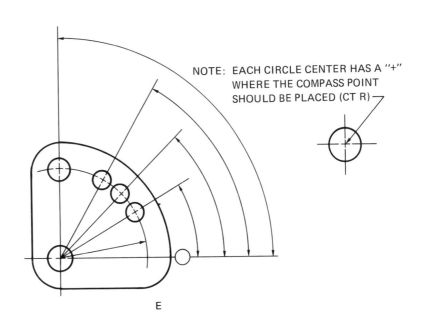

NOTE: EACH CIRCLE CENTER HAS A "+"
WHERE THE COMPASS POINT
SHOULD BE PLACED (CT R)

E

Fig. 11-28 Dimensioning hole centers on bolt circles

UNIT 12

ADVANCED DIMENSIONING

DIMENSIONING

Dimensioning is perhaps the most important part of the drafter's job. It is even more important than speed, neatness, and accuracy. A drawing can be of poor quality but, if dimensioned correctly, could be used. The ideal drawing, one that the beginning drafter should strive for, is a drawing that is neat, in exact scale, dimensioned correctly, and completed in the shortest amount of time.

A fully qualified drafter must know and fully understand how the part will be manufactured, what tolerances should be applied, how the part will function with other parts, and how to describe or illustrate the part on paper so there is no misinterpretation by the worker who will make the part.

No two parts can be made exactly alike, but they can be made within specific tolerances so they are interchangeable. Because of this fact, a tolerance system was devised which is a part of the national standard for dimensioning. It is the drafter's job to understand and use this system of tolerancing. All illustrations in this unit are in fractions and/or decimals. The metric system uses exactly the same process.

LIMITS AND TOLERANCE

Think of the signs along the interstate highways of our country. They tell motorists how fast or slow they are allowed to drive. If they go faster than posted, they could be fined for speeding. If they drive slower than posted, they could also be fined, figure 12-1.

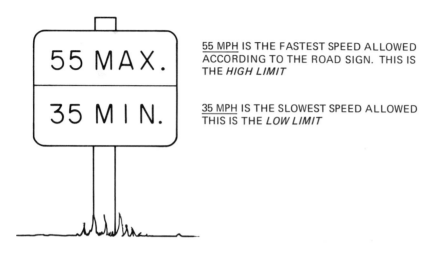

55 MPH IS THE FASTEST SPEED ALLOWED ACCORDING TO THE ROAD SIGN. THIS IS THE *HIGH LIMIT*

35 MPH IS THE SLOWEST SPEED ALLOWED THIS IS THE *LOW LIMIT*

Fig. 12-1 Driving tolerances

The *tolerance* in the example in figure 12-2 is the difference between the high limit and the low limit. Thus:

$$
\begin{array}{rl}
55 \text{ mph} & \text{high limit} \\
- 35 \text{ mph} & \text{low limit} \\
\hline
20 \text{ mph} & \text{accepted } tolerance
\end{array}
$$

The drafter must state the largest and the smallest size hole that is acceptable for a particular application or function.

In figure 12-2, .505 inch is the *largest* hole allowed, the high limit; .500 inch is the *smallest* hole allowed, the low limit. The tolerance in this example is the difference between the high limit and the low limit. Thus:

$$
\begin{array}{rl}
0.505'' & \text{diameter (high limit)} \\
- 0.500'' & \text{diameter (low limit)} \\
\hline
0.005'' & \text{design tolerance}
\end{array}
$$

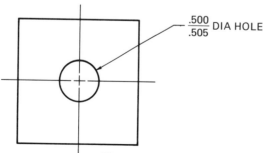

$\frac{.500}{.505}$ DIA HOLE

Fig. 12-2 Drawing tolerances

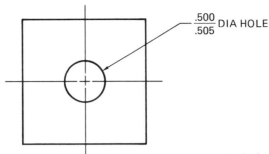

Fig. 12-3 Tolerance of a hole

Never call for closer or tighter limits than are necessary. Closer or tighter limits than necessary are more costly to produce. Try to design for the best function at minimum cost. Some things to consider when determining tolerances include:

- Length of time parts are engaged
- Speed (if any) mating parts will move or turn
- Lubrication
- Temperature & humidity

- Material used
- Estimated "life" required
- Capability of company to produce the tolerance
- Cost (very important)

In calling out a hole, place the smallest limit on top, figure 12-3. The theory for doing this is, if a machinist tries for the top figure (.500″) and the hole is too small, it can be redrilled larger. If it is drilled a little larger than the top limit (smallest), it will still fall within the given tolerance.

In calling out a shaft, place the largest limit on top, figure 12-4. The theory for doing this is, if a machinist tries for the top figure (.495″) and makes the shaft too large, it can be machined smaller. If it is machined a little smaller than the top limit (largest), it will still fall within the given tolerance. Note that various company standards differ, but for *all* problems done in this course use the standard presented here.

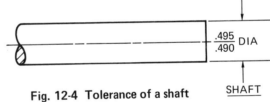

Fig. 12-4 Tolerance of a shaft

Maximum Material Condition

If an object with a hole size of .500-inch diameter was placed on a scale, it would weigh *more* than the same size object with a .505-inch diameter hole drilled in it, figure 12-5. The *maximum material condition* (MMC) is the smallest size the hole can be.

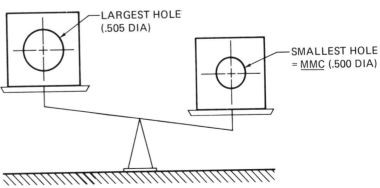

Fig. 12-5 MMC of a hole

If a shaft with a .495-inch diameter is placed on a scale, it would weigh *more* than a shaft with a .490-inch diameter. The *maximum material condition* (MMC) is the largest size the shaft can be, figure 12-6.

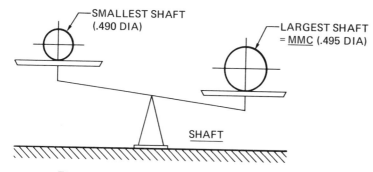

Fig. 12-6 Maximum material condition of a shaft

Thus, the *maximum material condition* is the condition where a feature of size contains the maximum amount of material within the stated limits of size.

ALLOWANCE OR CLEARANCE

Allowance is the intentional difference between the sizes of mating parts. Figure 12-7 shows how to find the minimum allowance, and figure 12-8 shows how to find the maximum allowance.

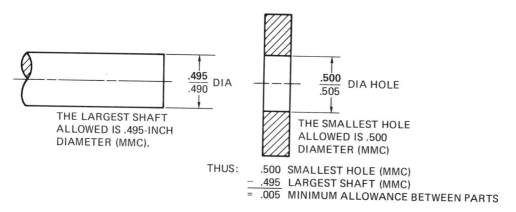

Fig. 12-7 Derivation of minimum allowance

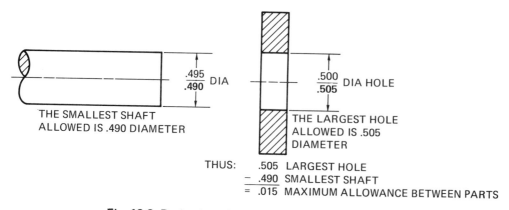

Fig. 12-8 Derivation of maximum allowance

Nominal Size

Nominal size is the designation used for general identification only. For example, a steel plate that is referred to as 1/4 inch (6) thick. If measured, however, it is actually more or less than 1/4 inch (6) thick.

Basic Size

Basic size is the size the drafter starts with before applying the required limits to it. There are two systems: the *basic hole system* and the *basic shaft system*. The basic hole system starts with the hole size (basic size) and adjusts the shaft size to fit. The latter system starts with the shaft size (basic size) and adjusts the hole size to fit. Because holes usually are made with standard tools (drills, reams, bores) it is best to use the basic hole system. The shafts can be made to most any size with little problem.

How to Determine Size Using the Basic Hole System

Use the basic hole system for all problems in this unit. Starting with the nominal size, determine the basic size (smallest hole), and calculate all sizes needed. Label each item as illustrated. Dimensions must be correct. Remember to dimension holes with the smallest limit on top, and the shaft with the largest limit on top, figure 12-9.

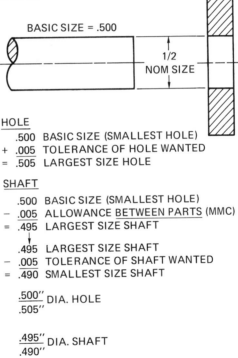

BASIC SIZE = .500

1/2
NOM SIZE

HOLE
.500 BASIC SIZE (SMALLEST HOLE)
+ .005 TOLERANCE OF HOLE WANTED
= .505 LARGEST SIZE HOLE

SHAFT
.500 BASIC SIZE (SMALLEST HOLE)
− .005 ALLOWANCE BETWEEN PARTS (MMC)
= .495 LARGEST SIZE SHAFT

.495 LARGEST SIZE SHAFT
− .005 TOLERANCE OF SHAFT WANTED
= .490 SMALLEST SIZE SHAFT

.500″
.505″ DIA. HOLE

.495″
.490″ DIA. SHAFT

Carefully transfer calculated values to the drawing:

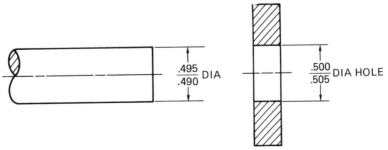

.495
.490 DIA

.500
.505 DIA HOLE

Fig. 12-9 Using basic hole system

KINDS OF FITS

A *fit* is a general term used to signify the range of tightness or looseness which results from the application of a specific combination of allowances and tolerances in mating parts. There are many kinds of fits. However, only a limited number will be discussed in this unit.

A *clearance fit* has limits of size that result in a clearance when mating parts are assembled. The parts can be assembled by hand because the hole is always larger than the shaft.

An *interference fit* has limits of size that result in an interference when mating parts are assembled. Parts must be pressed together because the hole is always smaller than the shaft.

In a *transition fit,* the limits of size result in either a clearance or interference when mating parts are assembled.

Remember: Dimensions of holes — smallest limit on top; dimensions of shafts — largest limit on top. A positive allowance = clearance. A negative allowance = interference.

Remember:

$$
\begin{array}{l}
\text{SMALLEST HOLE} - \textit{Lower limit } \text{hole} \\
\underline{+\quad\text{TOLERANCE}} \\
=\text{ LARGEST HOLE} - \textit{Upper limit } \text{hole}
\end{array}
$$

(Clearance fit) ◄———————— or ————————► *(Interference fit)*

SMALLEST HOLE	SMALLEST HOLE
<u>− ALLOWANCE</u>	<u>+ ALLOWANCE</u>
= LARGEST SHAFT − *Upper limit* shaft	= LARGEST SHAFT − *Upper limit* shaft
LARGEST SHAFT	LARGEST SHAFT
<u>− TOLERANCE</u>	<u>− TOLERANCE</u>
= SMALLEST SHAFT − *Lower limit* shaft	= SMALLEST SHAFT − *Lower limit* shaft

Keys and Slots

The limits, tolerances, and allowances for other close-fitting parts are determined the same way as for holes and shafts. In practice exercise 7-2, the slot takes the place of the hole, and the key takes the place of the shaft. All terms associated with hole and shaft apply.

TOLERANCING

Unilateral Tolerancing

Unilateral tolerancing means applying the tolerance in only one direction from the size actually wanted. Because parts cannot be made exactly alike, the drafter sometimes allows a larger tolerance than the exact size (dimension), but not smaller. This is called unilateral tolerance, figure 12-10.

Note: Maximum dimension on top, minimum dimension on bottom.

Fig. 12-10 Unilateral tolerance

Bilateral Tolerancing

Bilateral tolerancing means applying a portion of the tolerance in two directions from the size actually wanted. For example, a drafter knows that an exact 1.500-inch length cannot be made, so + .002 or –.002 inch from the size required is tolerated. This is called a bilateral tolerance, figure 12-11.

Fig. 12-11 Bilateral tolerance

Tolerance Allowance Chart

In drawing forms used by many companies, the title block includes a tolerance chart, such as the one in figure 12-12.

Tolerance unless otherwise spec.
Fractions +/–.015 *(1/64)*
.XX   +/–.015
.XXX   +/–.010
.XXXX +/–.0002

Fig. 12-12 Tolerance allowance chart

(Note: *"unless otherwise spec."* means any dimension that must be held closer will have limits directly on it.)

Using the tolerance chart in figure 12-12, any fractional dimension will automatically have a ± .015 inch (1/64) tolerance applied to it, any two-place decimal (.xx) will automatically have a ± .015-inch tolerance applied to it, etc. Using a 1/2-inch nominal size as reference:

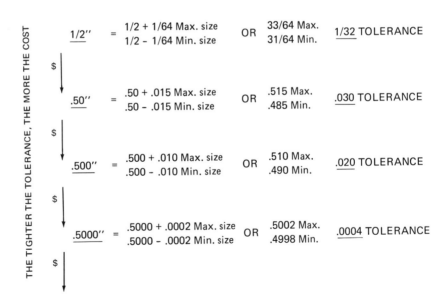

CALLOUTS FOR HOLES

Each manufacturing company has its own standard for calling out hole dimensions. Figure 12-13 illustrates the callout standard most frequently used in industry. The drafter does not specify on a drawing the process to be used but merely indicates the limits required of the hole. Specify the number of holes required, as illustrated, even if only one hole is made.

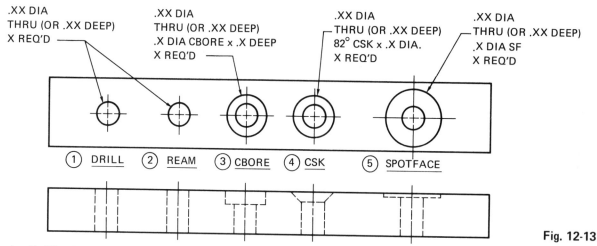

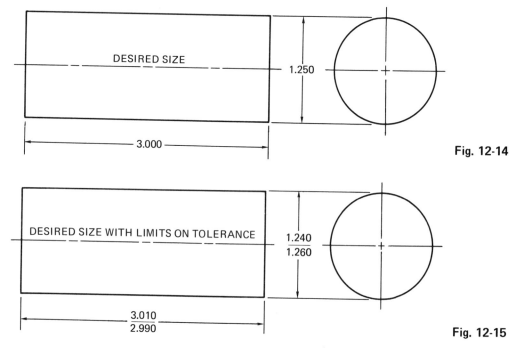

Fig. 12-13

1. Drilling is used most often to make holes. The *drilled hole* produced is acceptable unless a limit is indicated. If it is not clear on the drawing, the notation also indicates through holes or the depth of the hole, and the number required.

2. A *reamed hole* is called out the same as a drilled hole, except it will always have limits which can be made only with a reamer.

3. A *counterbored hole* must include hole diameter, diameter and depth of the counterbore and the number required.

4. A *countersunk hole* must include the hole diameter, angle of countersink (82-degree angle is standard), the diameter of countersink measured across the top of the countersunk hole, and the number required.

5. A *spotface* must include hole diameter, hole depth, spotface diameter, and number required. Do not call out spotface depth as it varies depending upon the surface irregularities. A spotface is made only to that depth necessary to produce a flat surface or a predetermined thickness recorded for a feature; i.e. a flange.

DRAWING LIMITS AND TOLERANCES

Figure 12-14 shows a drawing of the shaft size desired by the drafter. Figure 12-15 shows the same drawing with limits and tolerances added to the desired shaft size.

Fig. 12-14

Fig. 12-15

The drawing in figure 12-16 shows the size range in which the shaft must fall to be accepted by the inspection department. If the shaft was machined to a length less than 2.990 inch, it would be scrapped. If it was over 3.010 inches in length, it would be remachined to fit within given limits and tolerances.

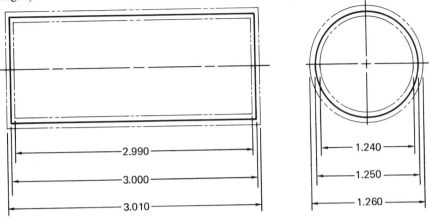

Fig. 12-16

The drawing in figure 12-17 shows the hole location desired by the drafter. Figure 12-18 is the same drawing with the hole location limits and tolerances added to desired hole location dimensions.

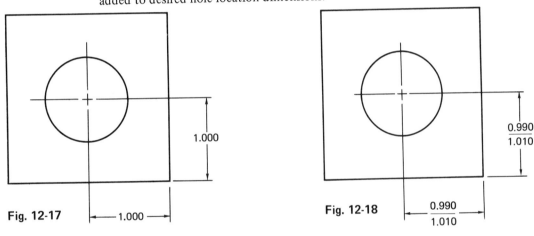

Fig. 12-17

Fig. 12-18

Figure 12-19 is the same drawing showing the effect of the hole location using acceptable limits and tolerances. If the hole location varies from the acceptable location limits and tolerances, the inspection department will scrap the part.

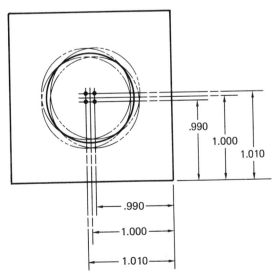

Fig. 12-19

TOLERANCE ERROR

A *buildup of tolerance* will result in tolerances much larger than those that are acceptable. Perform the following experiment to learn what tolerance buildup refers to.

1. Set a compass with a sharp lead to an exact opening of 11/16 inch.
2. Step off 10 equal 11/16-inch spaces starting at the point noted by "start" in figure 12-20.
3. Measure the total overall length stepped off.

START

Fig. 12-20

The length should be 6 7/8 inches (10 × 11/16"). If the answer is off it is because the original compass setting was inaccurate (a little long or short of 11/16 inch), and that inaccuracy was multiplied ten times. This type of error is called a buildup of tolerances.

The tolerance of one-place decimals in figure 12-21 is given as ± .015 inch. If the machinist makes the object using all long dimensions, as shown in the lower view, the object length totals 4.075 inches. This is considerably longer than the acceptable length of 4.015 inches. This occurred because of a tolerance buildup. The same situation would occur if the lower limits were used.

AS DRAWN AND DIMENSIONED

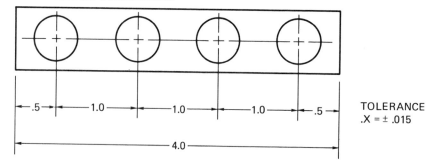

TOLERANCE
.X = ± .015

AS MANUFACTURED

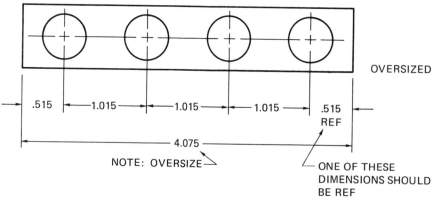

OVERSIZED

.515 ◄—1.015—►◄—1.015—►◄—1.015—► .515 REF

◄——————— 4.075 ———————►

NOTE: OVERSIZE

ONE OF THESE
DIMENSIONS SHOULD
BE REF

Fig. 12-21

Figure 12-22 shows how a tolerance buildup is avoided. If numerical control machines are used, there will not be any problem with incremental dimensioning in figure 12-21. This is becuase the inherent accuracy of the machines is closer than the specified limits.

Base Line Dimensioning

Base line dimensioning is the best solution to tolerance buildup. Using the left side (or any important edge, point, or center line as a feature) as a base or reference, project all dimensions from the same reference base.

AS DRAWN AND DIMENSIONED

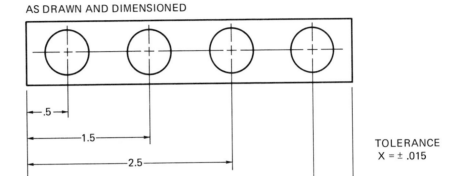

TOLERANCE
X = ± .015

SHOWN WITH ALL DIMENSIONS
MANUFACTURED OVERSIZE
WITHIN SPECIFICATION

AS MANUFACTURED

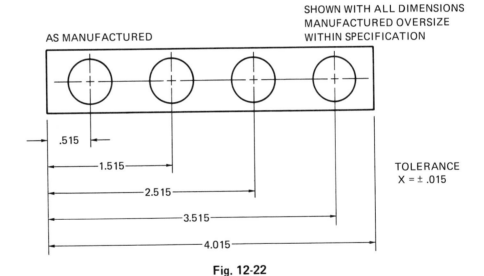

TOLERANCE
X = ± .015

Fig. 12-22

The drafter must consider the possibility one part could be made on the minus side of given tolerance and another could be made on the plus side of the factured, with all dimensions oversize but within specifications. The worse condition is either .015 inch oversize or .015 inch undersize. There is no chance of a tolerance buildup using this method of dimensioning.

The drafter must consider the possibility part 'A' could be made on the minus side of given tolerance and part 'B' could be made on the plus side of the given tolerance. The drafter should consider carefully where to place the dimensions, which dimensions should have close tolerances and which should be very loose, and, above all, how various mating parts go together. It is important that mating parts be dimensioned from the same base line or feature so that a buildup of tolerances cannot occur.

Study figure 12-23. Note that the holes in part 'A' must be large enough to allow for hole location tolerance, location tolerance, and size tolerance in part 'B'. Notice that even though an edge was used as a base line for dimensioning other holes, the base line could also have a *hole* from which other holes would be dimensioned. The important consideration is determining from the design which characteristic is the most important.

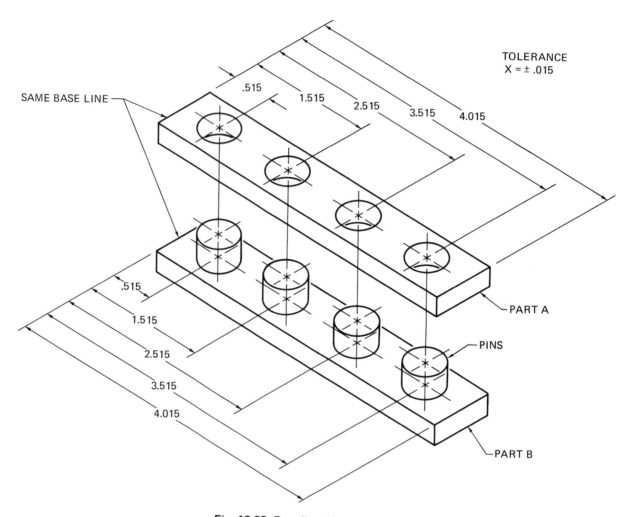

Fig. 12-23 Base line dimensioning

Other factors to consider in deciding the tolerances are:
- Length of time parts are engaged
- Speed at which mating parts move
- Load on part (s)
- Lubrication used (if necessary)
- Average temperature at which parts operate
- Humidity factor
- Materials from which parts are made
- Estimated "life" required of part
- Capability of machinery, tools, and machinist manufacturing part
- Cost

TYPES AND USES OF FITS

A standard system of fits is used to help manufacturers, for example, use fewer gauges and to provide the optimum design control. As a consequence, interchangeable parts are more readily achieved. Standard fits are noted in figure 12-24 with an explanation of where each is used. The drafter must choose as loose a fit as applicable to the design requirements. Unnecessarily close fits that are not required by the design are more costly to manufacture.

Sym.	Type of Fit	No.	Usage
RC	Running & Sliding	1	Accurate location — little or no "play."
		2	Accurate location—more clearance-turn or move at slow speed.
		3	Precision running at slow speed-light pressure-no temp. change.
		4	Accurate running at moderate speed-medium pressure-min. "play."
		5	High speed — heavy pressure.
		6	
		7	Free running — not accurate — large temperature changes.
		8	Loose — not accurate — low price.
		9	
LC	Locational Clearance	1	Normally stationary parts — easily assembled/disassembled.
		2	
		3	
		4	
		5	
		6	
		7	
		8	
		9	
		10	
		11	
LT	Transitional Locational	1	Close accuracy of mating parts — little or no interference.
		2	
		3	
		4	
		5	
		6	
		7	
LN	Locational Interference	2	Exact accuracy of mating parts.
		3	
FN	Force & Shrink	1	Light drive fit-permanent assembly — usually used with cast iron.
		2	Medium drive fit — or shrink fit — used for steel-small parts.
		3	Heavy drive fit — medium size parts — used for steel.
		4	Force fit or shrink fit in parts which can be highly stressed.
		5	

Fig. 12-24 Fits, kinds and uses

Choosing a Class of Fit

A drafter who is designing an assembly of 1 1/2-inch nominal size diameter shaft that must turn at a high speed under a heavy load or pressure must calculate tolerances on the hole and the shaft. In looking at figure 12-24, an RC–5 or RC–6 fit comes closest to the required specifications.

How to Figure Tolerance

Use A.S.M.E. #B4.1 chart, Appendix F.

1. The basic hole system is used in this method.

2. All limits on the chart are in 1000ths of an inch.

   For example:

   - 2.0 would mean 2 thousandth or .002 (move decimal point three places to the left).

   - 1.5 would mean 1 1/2 thousandths or .0015 (move decimal point three places to the left).

   - .8 would mean .8 thousandths or .0008 (move decimal point three places to the left).

3. Be sure to observe the plus (+) or minus (–) sign before each figure. All *clearance* fits must use minus (–) and all *interference* fits must use plus (+) to calculate the largest shaft — This is important!

4. Accuracy is a must. Check and double check all work.

   Example:

   Carefully do the following steps:

   1.) 1 1/2 inch (basic size) required (1.500")

   2.) Locate either RC-5 or RC-6 (RC-6 will be used as an example) on the Running/Sliding Chart, page 341.

   3.) Locate at the left of the chart, the range of size that 1 1/2 inch falls into (1.19 to 1.97).

   4.) Calculate the limits of the hole using figures from the chart for "Hole Tolerance"

   $$
   \begin{array}{ll}
   1.5000 & \text{smallest hole-} \\
          & \text{lower limit (Basic size)} \\
   + \ .0025 & \text{hole tolerance (from chart)} \\
   \hline
   1.5025 & \text{largest hole-upper limit}
   \end{array}
   $$

   5.) Calculate the limits of the shaft:

   $$
   \begin{array}{ll}
   1.5000 & \text{smallest hole (basic size)} \\
   - \ .0020 & \text{clearance allowance (from chart)*} \\
   \hline
   1.4980 & \text{largest shaft (upper limit, shaft)}
   \end{array}
   $$

   *Note — clearance allowances are *subtracted* (–) interference allowances are *added* (+)

   $$
   \begin{array}{ll}
   1.4980 & \text{largest shaft (upper limit, shaft)} \\
   - \ .0016 & \text{tolerance} \\
   \hline
   1.4974 & \text{smallest shaft (lower limit, shaft)}
   \end{array}
   $$

   6.) Check all math

This math procedure is the same as illustrated on page 168 using the basic "Hole System".

It should be noted that standard fits will apply in the majority of cases. However, when design requirements cannot be met with Standard Fits, use the same method for calculating the desired fit.

TOLERANCING DRILLED HOLES

The chart in Appendix G is a guide for determining the limits for a hole's dimensions when a particular size drill is used. Steps one through four outline how to use the charts. The example uses a #45 standard drill.

Step 1. Locate the letter or number of the drill at the left of the chart.

Step 2. Convert it to a decimal:

$$\#45 = .0820$$

Step 3. Find the upper limit:

$$.0820 + \text{tolerance of } .0043 = .0863$$

Step 4. Find the lower limit:

$$.0820 - \text{tolerance of } .0010 = .0810$$

The hole callout is: .0810 lower limit
                               .0863 upper limit

"The term drill is not used" — "diameter" is used — not the process to make the hole.

Example: A 1/2″ diameter hole would probably be *drilled.* A .5000 diameter hole would probably be *reamed. Both* should be called out as "diameter". It is up to the craftsperson to hold the tolerance, thus choose the correct method or process to arrive at the required size, and within tolerance.

Feature Control Symbol

Feature control symbols call out what planes (views) the tolerances will be in respect to. Sometimes a tolerance is tied to more than one plane. In this case the first, second, and auxiliary datum notes are used, figure 12-25.

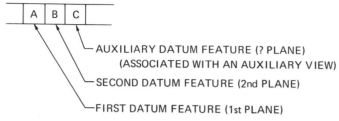

AUXILIARY DATUM FEATURE (? PLANE)
(ASSOCIATED WITH AN AUXILIARY VIEW)
SECOND DATUM FEATURE (2nd PLANE)
FIRST DATUM FEATURE (1st PLANE)

Fig. 12-25 Feature control symbol

If any surface does not have a geometric tolerance specified, the form is allowed to vary within the given limits of size.

Flatness

Flatness means that the entire surface must lie between two parallel planes that cannot be more than a specified tolerance apart. The symbol for flatness is a *parallelogram*, figure 12-26. Note that flatness applies to flat surfaces only.

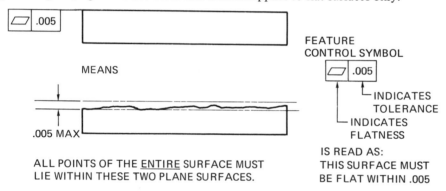

MEANS

.005 MAX

ALL POINTS OF THE ENTIRE SURFACE MUST
LIE WITHIN THESE TWO PLANE SURFACES.

FEATURE
CONTROL SYMBOL

.005

INDICATES
TOLERANCE
INDICATES
FLATNESS

IS READ AS:
THIS SURFACE MUST
BE FLAT WITHIN .005

Fig. 12-26 Geometric tolerancing, flatness

Straightness (−)

Straightness means that the entire surface must be straight within a given limit. Straightness applies to cylinder or cone surfaces only, figure 12-27.

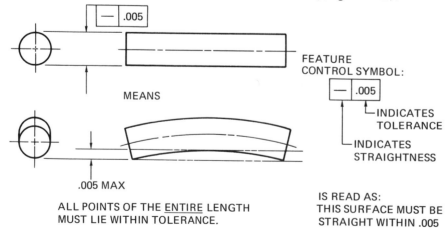

MEANS

.005 MAX

ALL POINTS OF THE ENTIRE LENGTH
MUST LIE WITHIN TOLERANCE.

FEATURE
CONTROL SYMBOL:

.005

INDICATES
TOLERANCE
INDICATES
STRAIGHTNESS

IS READ AS:
THIS SURFACE MUST BE
STRAIGHT WITHIN .005

Fig. 12-27 Geometric tolerancing, straightness

Cylindricity

Cylindricity is a combination of straightness, roundness, and parallelism. A cylindricity tolerance refers to all three of these properties. Note that both circles, figure 12-28, are about the same axis.

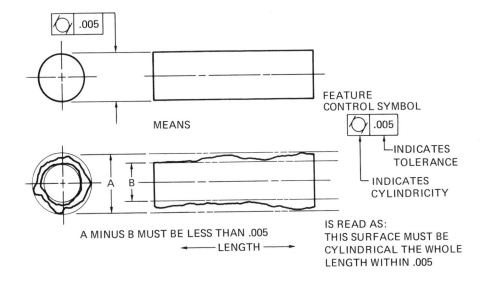

Fig. 12-28 Geometric tolerancing, cylindricity

Roundness

Roundness refers to the circular form of an object. A roundness tolerance describes the range within which the radius of a circle may vary. One of the features of a cylindrical object is that it has a round cross section, figure 12-29.

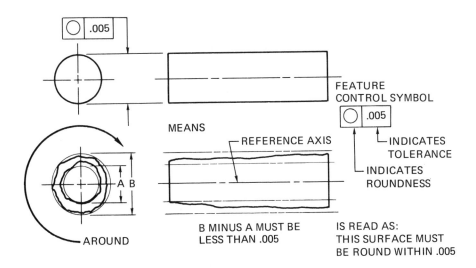

Fig. 12-29 Geometric tolerancing, roundness

Parallelism ∥

Parallelism means that an entire surface must lie between two parallel planes that cannot be more than a specified tolerance apart, parallel to and in relation to a given surface or datum. In figure 12-30, the bottom surface is datum "A."

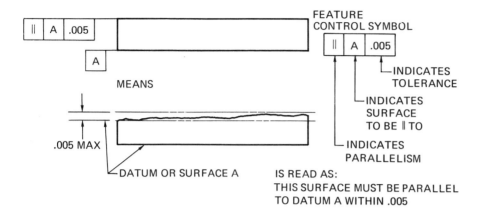

Fig. 12-30 Geometric tolerancing, parallelism

Remember a datum could be a surface, point, center line or any feature on the object.

Perpendicularity ⊥

Perpendicularity means that an entire surface must lie between two parallel planes that cannot be more than a specified tolerance apart, perpendicular to and in relation to a given surface or datum. The upright surface in figure 12-31 is datum "A."

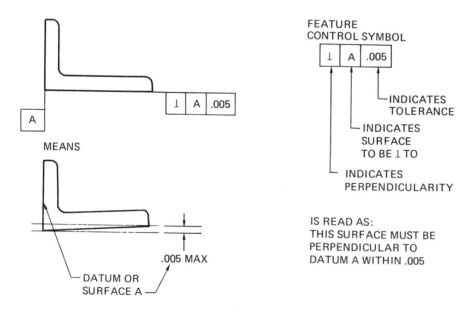

Fig. 12-31 Geometric tolerancing, perpendicularity

Angularity ∠

Angularity means that the entire surface must lie between two parallel planes which are at the true angle in relation to a specified surface or datum (more than one datum can be used) with a specified tolerance, figure 12-32.

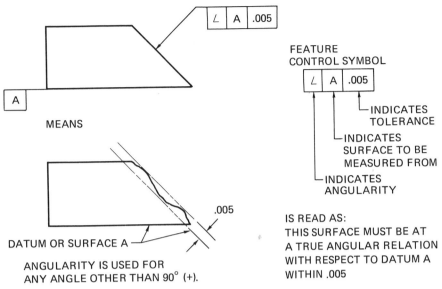

Fig. 12-32 Geometric tolerancing, angularity

Concentricity ◎

Concentricity means that the axis of one feature or diameter must lie within a cylindrical tolerance zone, which is concentric to the datum axis of another feature or diameter within a specified tolerance, figure 12-33.

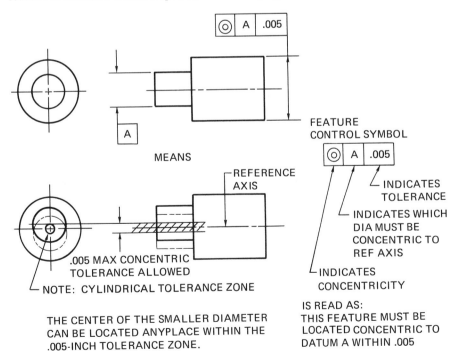

Fig. 12-33 Geometric tolerancing, concentricity

Symmetry

Symmetry means that the entire feature must lie between two parallel planes that cannot be more than a specified tolerance apart symmetrically and in regards to a surface(s) or datum. The specified tolerance in this case must be *equally spaced between the datum,* or surface(s), figure 12-34.

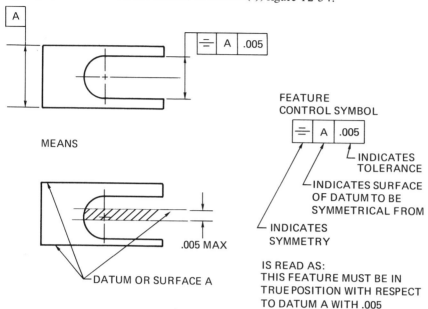

MEANS

.005 MAX

DATUM OR SURFACE A

FEATURE
CONTROL SYMBOL

⎓ | A | .005

INDICATES
TOLERANCE

INDICATES SURFACE
OF DATUM TO BE
SYMMETRICAL FROM

INDICATES
SYMMETRY

IS READ AS:
THIS FEATURE MUST BE IN
TRUE POSITION WITH RESPECT
TO DATUM A WITH .005

Fig. 12-34 Geometric tolerancing, symmetry

Modifiers

Sometimes a *modifier* is added to the feature control symbol:

- Ⓢ means "regardless of feature size" (rfs). This reads as: parallel to surface A, regardless of feature size and within .005 inch.
- Ⓜ means "maximum material condition" (MMC). This reads as: parallel to surface A when it is at maximum material condition only and within .005 inch.

| ‖ | A | Ⓢ | .005 |

| ‖ | A | Ⓜ | .005 |

Total Indicator Reading (TIR)

The note TIR on a drawing means *total indicator reading.* This means the part to be checked must be round the full 360 degrees to within .005 inch. The part is set up so it can rotate about a fixed center line. An indicator is mounted above it and set on 0. The part is then rotated and the indicator must not move more than a total of .005 inch in both directions. If it does, it is not within tolerance, figure 12-35.

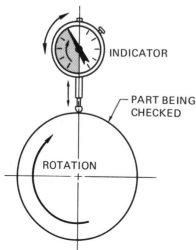

INDICATOR

PART BEING
CHECKED

ROTATION

Fig. 12-35 Dial indicator measuring tolerance

TRUE POSITION

Using the normal bilateral system of dimensioning, a hole with limits of + or − .015-inch is dimensioned as in figure 12-36.

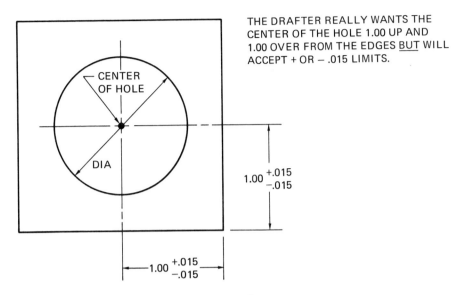

Fig. 12-36

If the above limits are accepted, the *tolerance zone* forms a square area that, in effect, establishes boundaries in which the center of the circle is located. In figure 12-37, the .030-inch tolerance zone extends the full length of the hole.

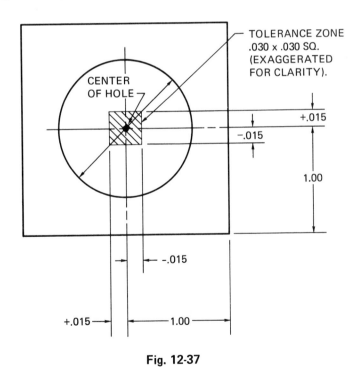

Fig. 12-37

Example 1. The four examples in figure 12-38, illustrate the worst conditions, maximum limits. Vertically and horizontally, the center of the hole can be from the design size and still be within limits.

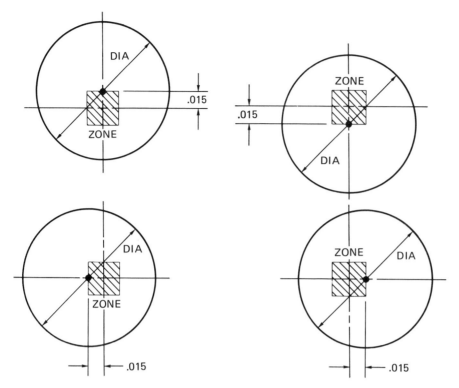

Fig. 12-38

Example 2. The four examples in figure 12-39 show the worst conditions, maximum limits. Diagonally across corners, the center of the hole can be from the design size and still be within limits.

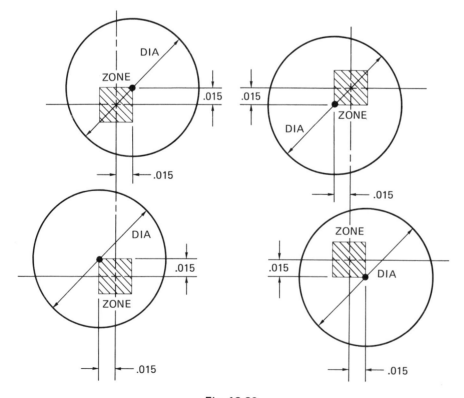

Fig. 12-39

In example 1, figure 12-38, the maximum distance from the design size still within limits is .015 inch, figure 12-40.

In example 2, figure 12-39, the maximum distance from the design size is much more, figure 12-39.

Using simple trigonometry, it is possible to calculate the exact diagonal distance from the center of the hole (.021-inch radii). In effect, there are two sizes or limits used in locating the center of the hole, ± .015 inch one way and ± .021 inch the other way, using *bilateral tolerancing*, figure 12-42.

If .021 inch is within tolerance, allow a .021-inch radius all around. A circle has 57 percent more area than a square. This will reduce scrap, reduce inspection time, reduce cost, and, in effect, still allow the exact same limits as the bilateral system (across corners), figure 12-43.

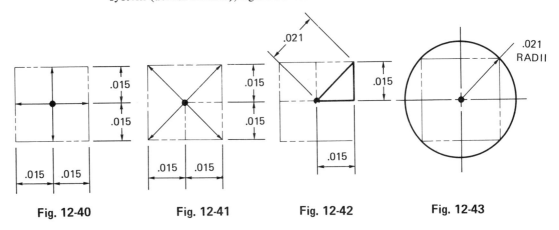

| Fig. 12-40 | Fig. 12-41 | Fig. 12-42 | Fig. 12-43 |

Simply stated, *true position* (TP) *enlarges* the tolerance zone. The symbol for true position is: ⊕

Comparing the bilateral tolerancing system (figure 12-44) with the true positioning system, it is easy to see the tolerance zone using true positioning is much larger, figure 12-45. This will result in fewer rejections of parts and lower costs. There is a 57% larger tolerance zone using true positioning.

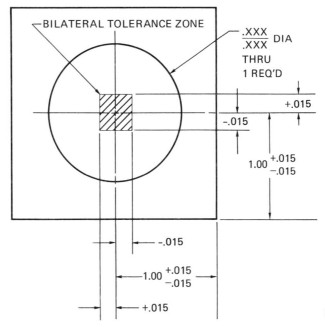

Fig. 12-44

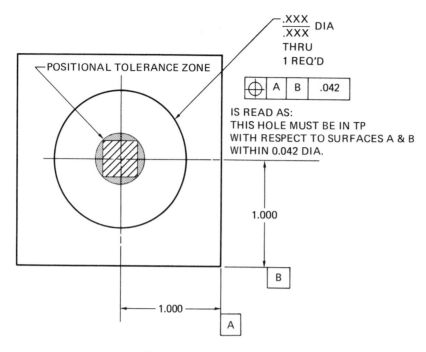

Fig. 12-45

Figure 12-46 is a list of most of the standard symbols used in engineering.

SYMBOLS

▱	FLATNESS
—	STRAIGHTNESS
⌭	CYLINDRICITY
○	ROUNDNESS
‖	PARALLELISM
⊥	PERPENDICULARITY
∠	ANGULARITY
◎	CONCENTRICITY
=	SYMMETRY
Ⓢ	REGARDLESS OF FEATURE SIZE
Ⓜ	MAXIMUM MATERIAL CONDITION
⊕	TRUE POSITION

Fig. 12-46 Symbol Chart

UNIT 13

MANUFACTURING PROCESSES

PROCESSES

The draftsperson must know and have a basic understanding of fundamental shop processes and terms in order to intelligently design dimensions and provide necessary information for material and finishes for components.

METALLURGY

A drafter must know about the behavior, characteristics, and properties of metals. This unit gives a very general working knowledge of metals, but much more on-the-job study must be done by the drafter.

Pure metals by themselves are usually too soft and weak to be used for machine parts. Thus, alloys are used. An *alloy* is simply a mixture of metals and chemical elements.

Materials must be carefully chosen to give the best working life of the part to be made and still be in line cost-wise with competition. In industry, most companies have one or more metallurgists who works with the engineering department to assist in the selection of correct metal or alloy for the design and function of each machine part. *Metallurgy* is the art and science of separating metals from their ores and preparing them for use.

CHARACTERISTICS OF METALS AND ALLOYS

The composition of metal and various chemical elements regulates the mechanical, chemical, and electrical properties of that metal. The following terms describe certain characteristics and capabilities associated with metals and alloys:

Strength is the ability to resist deformation.

Plasticity is the ability to withstand deformation without breaking. Usually hardened metals have strength but are very low in plasticity. They are brittle.

Ductility describes how well a material can be drawn out. This is an especially important characteristic for wire drawing and metal shape forming.

Malleability is the ability of a metal to be shaped by hammering or rolling.
Elasticity is the ability of metal to be stretched and then return to its original size.

Brittleness is a characteristic of metal to break with little deformation.

Toughness describes a metal that has high strength and malleability.

Fatigue limit is the stress, measured in pounds per square inch, at which a metal will break after a certain number of repeated applications of a load has been applied.

Conductivity describes how well a metal transmits electricity or heat.

Corrosion resistance describes how well a metal resists rust. Note that *rust* adds weight, reduces strength, and ruins the overall appearance of a metal.

HEAT TREATMENT

Heat treatment of metals and alloys provides certain desirable properties. Listed are a few basic terms associated with heating treatment:

Aging is a process that takes place slowly at room temperature.

Quenching is the process of cooling a hot metal in water or oil. In special cases it can be quenched in sand, lime, or asbestos rather than liquid in order to slow down the cooling process.

Tempering or *"drawing"* is the process of reheating and cooling by air. Tempering increases toughness, decreases hardness, relieves stress, and removes some brittleness.

Annealing is the process of softening a hardened metal so it can be shaped or machined. Annealing also removes internal stresses which cause warping and other deformations.

Case hardening hardens only the outer layer of the material. In this process, the outer layer of the metal absorbs carbon or nitrogen, thus hardening the outer layer.

Hardness testing is done by two methods: the Brinell hardness test or the Rockwell hardness test. The *Brinell method* uses a hardened steel ball that is pressed into the surface of the metal under a given pressure or load. The depth the ball goes into the metal surface is measured by a microscope and is converted into a hardness reading. The *Rockwell method* is very much the same except the hardness value is read directly from a scale attached to the tester.

DRAWING NOTATIONS

It is important that any part to be hardened is noted on the drawing. This notation must include the material, the heat-treating process, and the hardness test method and number.

TYPES OF METALS

There are two classifications of metal: (1) *ferrous*, or those that contain iron; and (2) *nonferrous*, or those that do not contain iron.

Ferrous Metals

Cast iron is widely used for machine parts. It is relatively inexpensive and easily cast into most any shape. It is a hard metal, strong, and has a good wearability. It responds very easily to almost all heat treating processes but tends to be brittle. *Malleable iron* is used where parts are subject to shock.

Steel is an alloy composed of iron and other chemical elements. It is important to remember that carbon content in steel regulates the properties of various types of steel. There are four classes of steel: carbon, alloy, stainless, and tool. A standard system of designating steel has been established by the American Iron and Steel Institute (A.I.S.I.) and the Society of Automotive Engineers (S.A.E.) to describe the type of steel to be used. The drafter must indicate the A.I.S.I.–S.A.E. numbers as indicated in figure 13-1. SAE1010 steel, for example indicates carbon steel with approximately 0.10% carbon.

PROPERTIES, GRADE NUMBERS & USAGES			
Class of Steel	*Grade Number	Properties	Uses
Carbon - Mild 0.3% carbon	10xx	Tough - Less Strength	Rivets - Hooks - Chains - Shafts - Pressed Steel Products
Carbon - Medium 0.3% to 0.6% carbon	10xx	Tough & Strong	Gears - Shafts - Studs - Various Machine Parts
Carbon - Hard 1.6% to 1.7%	10xx	Less Tough - Much Harder	Drills - Knives - Saws
Nickel	20xx	Tough & Strong	Axles - Connecting Rods - Crank Shafts
Nickel Chromium	30xx	Tough & Strong	Rings Gears - Shafts - Piston Pins - Bolts - Studs - Screws
Molybdenum	40xx	Very Strong	Forgings - Shafts - Gears - Cams
Chromium	50xx	Hard W/Strength & Toughness	Ball Bearings - Roller Bearing - Springs - Gears - Shafts
Chromium Vanadium	60xx	Hard & Strong	Shafts - Axles -Gears - Dies - Punches - Drills
Chromium Nickel Stainless	60xx	Rust Resistance	Food Containers - Medical/Dental Surgical Instruments
Silicon - Manganese	90xx	Springiness	Large Springs

*The first two numbers indicate type of steel, the last two numbers indicate the approx. average carbon content — 1010 steel indicates, carbon steel w/approx. 0.10% carbon.

Fig. 13-1 Properties, grade numbers, and usage of steel alloys

Nonferrous Metals

Copper is soft, tough, and ductile. It is a good conductor of both electricity and heat.

Brass is an alloy of copper (copper/zinc) and very workable, tough, and ductile.

Bronze is another alloy of copper (copper/tin). It is a serviceable, strong, and tough metal.

Aluminum is very malleable, ductile, and a good conductor of electricity and heat. It is very light in weight. Aluminum cannot be heat treated; thus, to increase its hardness, other alloys and elements must be added.

Magnesium is perhaps the lightest metal used today. It is a good conductor of electricity and heat, nonmagnetic, easily machined, but highly inflammable while machining.

SHAPES OF METALS

Metals are purchased from the manufacturer in standard shapes and sizes. The drafter must use the correct callouts when indicating what material is to be used, figure 13-2.

Extrusion is a method of forming very odd or special shapes, similar to squeezing toothpaste from a tube. The round opening is like a die (the required shape) and the toothpaste represents the metal to be shaped.

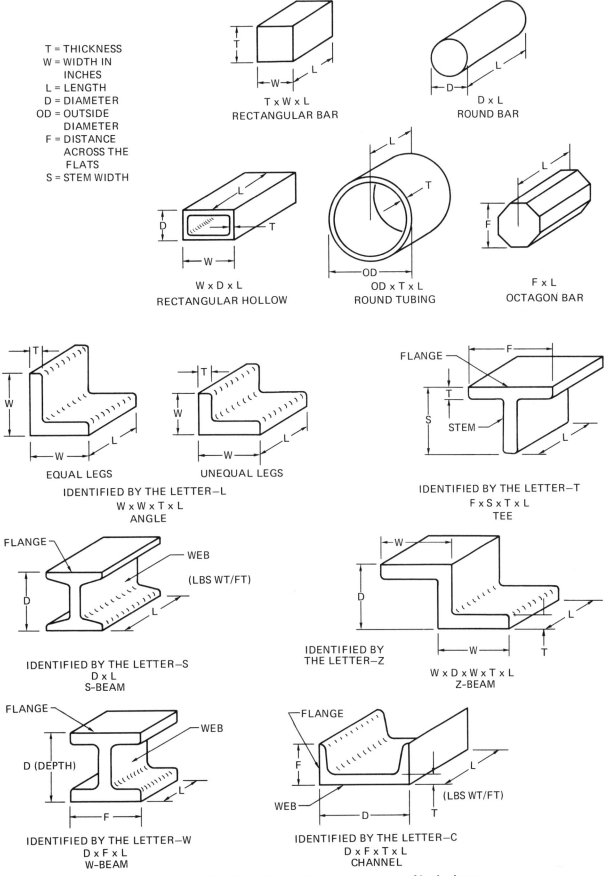

T = THICKNESS
W = WIDTH IN INCHES
L = LENGTH
D = DIAMETER
OD = OUTSIDE DIAMETER
F = DISTANCE ACROSS THE FLATS
S = STEM WIDTH

T x W x L
RECTANGULAR BAR

D x L
ROUND BAR

W x D x L
RECTANGULAR HOLLOW

OD x T x L
ROUND TUBING

F x L
OCTAGON BAR

EQUAL LEGS

UNEQUAL LEGS

IDENTIFIED BY THE LETTER—L
W x W x T x L
ANGLE

FLANGE
STEM

IDENTIFIED BY THE LETTER—T
F x S x T x L
TEE

FLANGE
WEB
(LBS WT/FT)

IDENTIFIED BY THE LETTER—S
D x L
S-BEAM

IDENTIFIED BY THE LETTER—Z
W x D x W x T x L
Z-BEAM

FLANGE
WEB
D (DEPTH)

IDENTIFIED BY THE LETTER—W
D x F x L
W-BEAM

FLANGE
WEB
(LBS WT/FT)

IDENTIFIED BY THE LETTER—C
D x F x T x L
CHANNEL

Fig. 13-2 Designating measurements of basic shapes

WEIGHTS OF MATERIALS

Figure 13-3 indicates the average weight per cubic foot of certain materials.

WEIGHTS OF MATERIALS

Material	Avg. Lbs. per Cu. Ft.	Avg. Kg. per Cu. Metre	Material	Avg. Lbs. per Cu. Ft.	Avg. Kg. per Cu. Metre
Aluminum	167.1	2676	Mahogany, Honduras, dry	35	564
Brass, cast	519	8296	Manganese	465	7448
Brass, rolled	527	8437	Masonry, granite or		
Brick, common and			limestone	165	2648
hard	125	2012	Nickel, rolled	541	8649
Bronze, copper 8, tin 1	546	8754	Oak, live, perfectly dry		
Cement, Portland, 376 lbs.			.88 to 1.02	59.3	953
net per bbl	110–115	1765–1836	Pine, white, perfectly dry	25	388
Concrete, conglomerate,			Pine, yellow, southern dry	45	706
with Portland cement	150	2400	Plastics, molded	74–137	1200–2187
Copper, cast	542	8684	Rubber, manufactured	95	1518
Copper, rolled	555	8896	Slate, granulated	95	1518
Fibre, hard	87	1377	Snow, freshly fallen	5–15	70–247
Fir, Douglas	31	494	Spruce, dry	29	459
Glass, window or plate	162	2577	Steel	489.6	7837
Gravel, round	100–125	1586–2012	Tin, cast	459	7342
Iron, cast	450	7201	Walnut, black, perfectly dry	38	600
Iron, wrought	480	7695	Water, distilled or pure rain	62.4	988
Lead, commercial	710	11,367	Zinc or spelter, cast	443	7095

Fig. 13-3 Average weights of materials

MACHINE TOOL OPERATIONS

The drafter must include all information required so a skilled machinist can manufacture a finished part from raw stock. Basic manufacturing processes shape raw stock by:

- *Cutting* into shape
- *Molding* into shape by casting or machine press
- *Pounding* into shape by forging
- *Forcing* into shape by bending
- *Fabricating* into shape, using parts manufactured from a combination of the above processes, by welding, riveting, screwing, or nailing the parts together

Machines that cut metal are called *machine tools.* There are over 400 kinds of machine tools, each designed to do a specific operation. All machine tool operations can be divided into five basic processes: Drilling, turning, planing, milling, and grinding.

A drafter does not have to be a machinist but should have a basic knowledge of machine tool operations in order to dimension drawings and "talk the language" of the skilled machinest. It is recommended that every mechanical drafting student visit a machine shop and study the five machine tool operations. A serious student should also try to take a mini-course in basic machine shop.

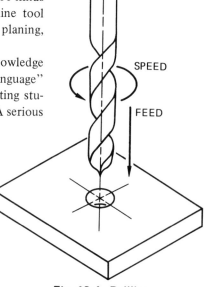

SPEED

FEED

Fig. 13-4 Drilling

Drilling

Drilling is probably the most basic of all machine tool operations. The process is done with a rotating tool called a *drill*, figure 13-4. The drill is held by a *chuck* and is rotated and fed into the part to be drilled. Holes are drilled before boring, reaming, countersinking, or counterboring operations can be completed.

Turning

Turning is the process of rotating the part that is to be machined and carefully pressing a cutting tool against it as it rotates, figure 13-5. A *lathe* is used for turning down stock and can be used for other machine operations such as drilling, boring, threading, cutting, milling, grinding, and knurling. A *turret lathe* is a lathe with a six-sided tool holder, called a *turret,* to which various cutting tools are attached. This attachment enables the lathe to do many operations without resetting the tools once each of the tools in the turret has been set.

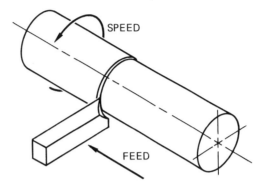

Fig. 13-5 Turning

Planing

Planing is the process of shaving material from raw stock very much like a carpenter does with a simple hand plane. The major difference is that the plane or cutting edge is stationary, and the part that is to be planed is moved back and forth, figure 13-6. A *shaper* is like a planer except, in shaping, the plane or cutting edge moves and the part is stationary. A *broach,* which is generally used to cut key slots and similar configurations, falls into the category of a planer in the way it operates. These tools can be worked horizontally, vertically, or angularly.

Grinding

Grinding is a machine operation where the part is brought into contact with a rotating abrasive wheel, figure 13-7. With this process it is possible to obtain very close, precise tolerances. Grinders that finish round parts are called *cylindrical grinders.* Those that grind flat parts are called *surface grinders.* Those that grind holes are called *internal grinders.*

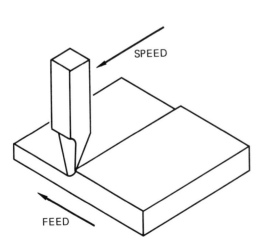

Fig. 13-6 Planing

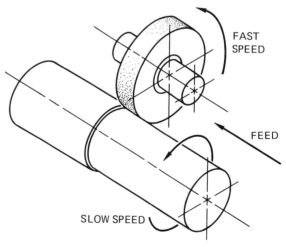

Fig. 13-7 Grinding

Milling

Milling is the process of bringing the part into contact with a rotating cutting tool having many edges, figure 13-8. The shape of the cutting edges are similar to a woodworking circular saw blade except they are usually wider and have a greater variety of shapes, figures 13-9 and 13-10. Milling machines produce cuts that are flat, round, sharp, and a combination of these shapes. Common milling machine processes are cutting slots and grooves, cutting gear teeth, making threads, boring holes, and rounding corners of parts.

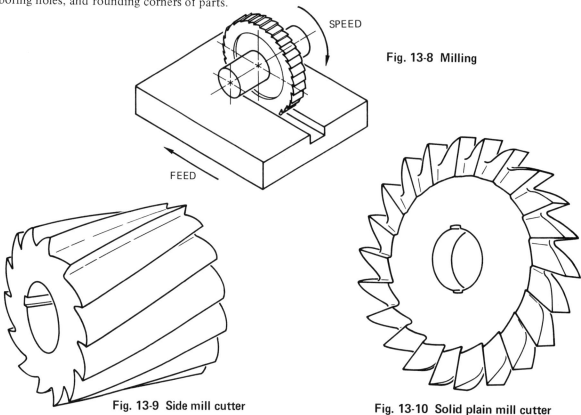

Fig. 13-8 Milling

Fig. 13-9 Side mill cutter

Fig. 13-10 Solid plain mill cutter

CASTINGS

Casting is the process of forming metal parts to rough size and shape by pouring molten metal into a mold. This process is similar to the way ice cubes are formed by pouring water into a tray and freezing it.

There are many forms of casting, varying in techniques and precision. To explain casting, one basic method — sand casting — is illustrated. Figure 13-11 shows the object to be cast.

The patternmaker constructs the pattern of the object to be cast. If the object, such as a bookend, has one flat side, the pattern is a one-piece pattern. If the object to be cast is round, a two-piece *split pattern* is used, figure 13-12.

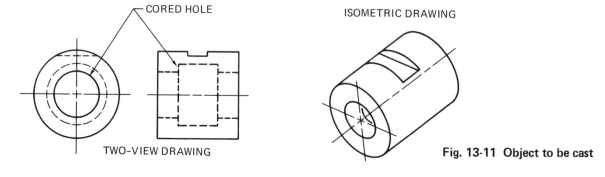

TWO-VIEW DRAWING

Fig. 13-11 Object to be cast

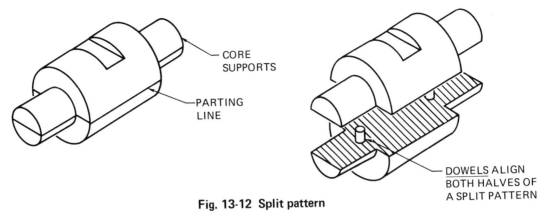

Fig. 13-12 Split pattern

If the object has a large hole through it, a *core support* is located inside the mold, figure 13-13. This prevents molten metal from solidifying in the hole during the casting process. When the core is removed, a rough cored hole remains which is later machined to the correct size. No cores are used on small holes as they are simply drilled into the finished casting.

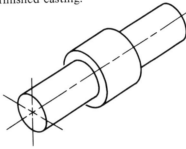

Fig. 13-13 Core made from baked sand and held together with a bonding agent

The *flask* is a hollow box with no top or bottom that holds the sand and the mold, figure 13-14. The *cope* is the top half of the flask and the *drag* is the lower half of the flask. A *socket* aligns the cope and drag. The flask is placed on a *molding board*.

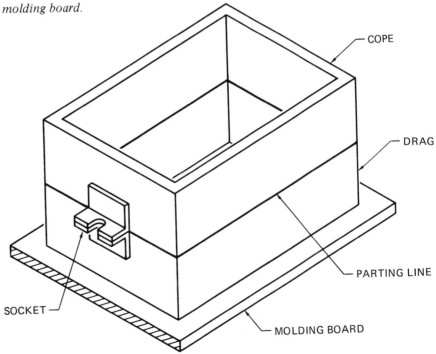

Fig. 13-14 A flask

How to Make a Metal Casting

The following description of a sand casting is a general description only and is not intended to be used in actual casting processes.

Step 1. The lower half of the split pattern is placed on the molding board. The drag is centered around the pattern, figure 13-15.

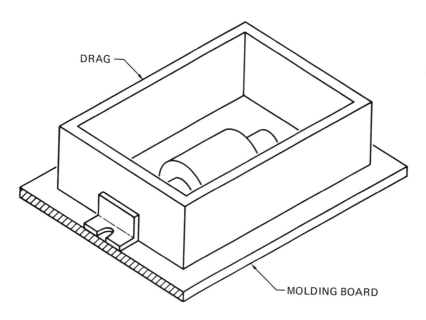

Fig. 13-15 Step 1

Step 2. The drag is filled with sifted sand which is packed firmly around the split pattern and leveled off. Another molding board is placed on top to hold the sand in place, and the drag is turned over, figure 13-16.

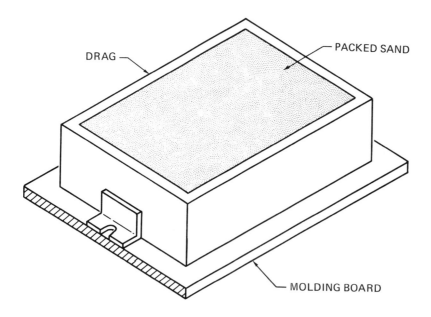

Fig. 13-16 Step 2

Step 3. The molding board is removed from the top to expose the pattern, figure 13-17.

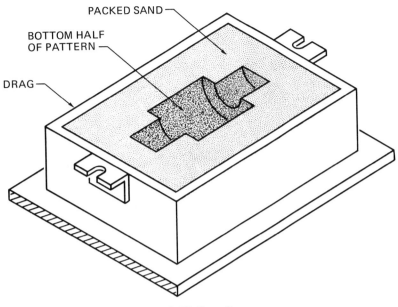

Fig. 13-17 Step 3

Step 4. The dowel pins are put in place and the top of the split pattern is positioned on the lower half, figure 13-18. The cope is then placed on the drag, locked into position, and filled with sand. The sand is tightly packed around the pattern.

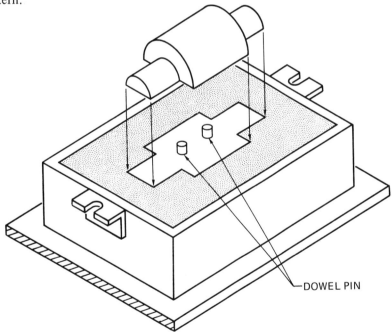

Fig. 13-18 Step 4

Step 5. Molten metal is poured into the mold through the *sprue hole.* The gases escape through the *riser hole* during the casting process. These holes are made while the pattern is in place so that the sand from the cope will not be forced into the hollow mold. An alternate method is to locate the sprue and riser holes to the left and right of the pattern and, with the flask apart, cut a groove leading from them to the mold, figure 13-19.

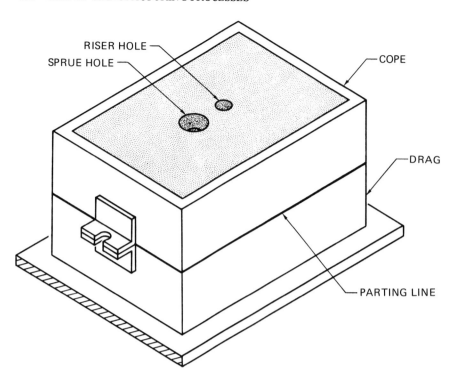

RISER HOLE

SPRUE HOLE

COPE

DRAG

PARTING LINE

Fig. 13-19 Step 5

Step 6. After cutting the sprue and riser holes, the cope is carefully removed from the drag and set aside, and the pattern removed from the drag. This leaves the top half of the mold in the cope and the bottom half of the mold in the drag. Round patterns will lift easily from the sand. Flat patterns tend to stick and can be damaged when removed. To prevent this, flat patterns are tapered on their sides. This taper is called a *draft*. The angle of draft is shown on a casting drawing and varies with the kind of material being cast. Sharp corners on flat patterns are also rounded off slightly for the same reason.

The core is put in place and the cope returned to the drag and locked into position. The casting is now ready to be poured. After the molten metal has solidified, the casting is removed from the sand and the core is removed from the casting. The casting is now ready to be machined, figure 13-20.

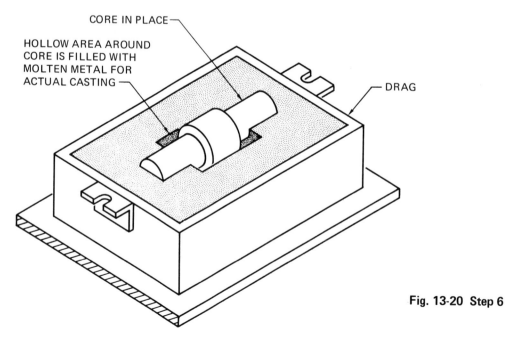

CORE IN PLACE

HOLLOW AREA AROUND
CORE IS FILLED WITH
MOLTEN METAL FOR
ACTUAL CASTING

DRAG

Fig. 13-20 Step 6

Section View of a Casting

A complex casting is drawn in two sections: One shows the pattern and the other shows the machining of the casting.

Figure 13-21 is a section view of a complete flask with the core in place. The molten metal is poured into the funnel-shaped sprue hole until the metal fills the cavity and comes out the riser. The riser allows air to escape while the mold is filled and feeds the casting while it cools.

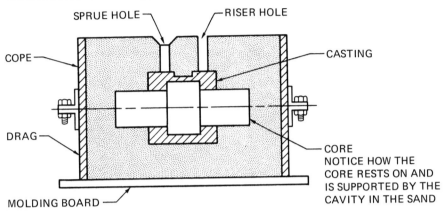

Fig. 13-21 Section view of mold ready to be cast

The sand core is easily broken up and removed, leaving a cavity inside the casting. When the casting is removed from the sand, the sprue and riser are still attached, figure 13-22.

The sprue and riser are easily removed by breaking or cutting them off, figure 13-23. They are then hand-ground. A sand casting is rough and the critical surfaces must be smoothed and machined to exact sizes.

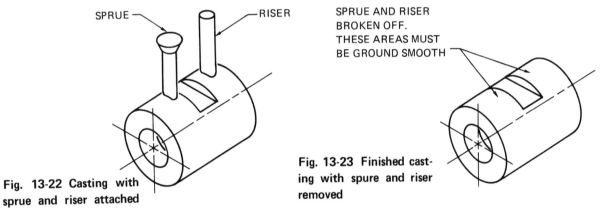

Fig. 13-22 Casting with sprue and riser attached

Fig. 13-23 Finished casting with spure and riser removed

Rounds and Fillets

Any part that is formed by casting should be designed with rounds and fillets, figure 13-24. *Rounds* are merely rounded outside corners. *Fillets* are rounded inside corners.

Rounds and fillets are used for three reasons:

1. Greater strength for inside corners (fillets)
2. Safer to handle; no sharp corners (rounds)
3. Gives the finished casting a much better appearance (rounds/fillets)

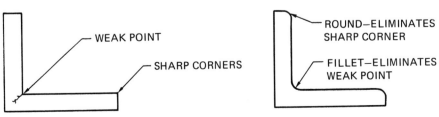

Fig. 13-24 Rounds and fillets

Bosses, Pads, and Machining Lugs

Bosses and *pads* serve the same function. They are raised surfaces that are machined to provide a smooth surface for mating parts. This method of designing saves material and machining time.

A *boss* is a round, raised surface. A *pad* can be any shape raised surface. Usually the bottom surface of a part is machined to provide a smooth, solid surface to support the part. The rest of the sand casting's surface is very rough. Proper use of rounds and fillets are illustrated in figure 13-25.

A *machining lug* is an extension of a surface to be machined. It is used for holding the casting because of its shape while machining. It too is removed after machining.

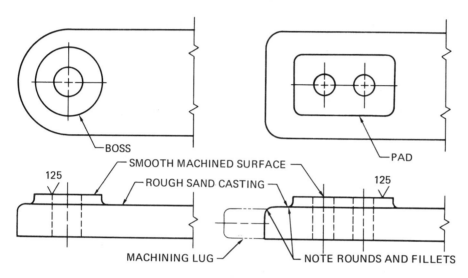

Fig. 13-25 Boss, pad, and machining lug

Ribs and Webs

Ribs and webs are similar and often confused. Think of a *rib* as a member that supports other members. A *web* simply connects various members together. Study figure 13-26. In designing castings, the general rule is to try to make all ribs and webs the same thickness. Otherwise, when the molten metal cools, thicker members cool last and tend to create warping and internal stresses.

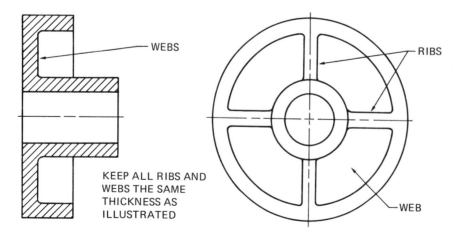

Fig. 13-26 Ribs and webs

Lug

A *lug,* sometimes referred to as an *ear,* is an extension added to the main part of the object, figure 13-27. It usually holds parts together. The thickness of the lug should be approximately the same thickness as the round body thickness.

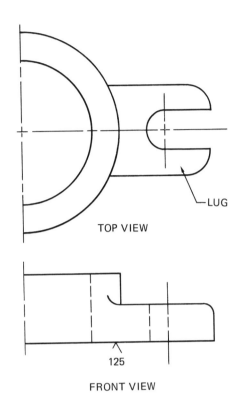

TOP VIEW

FRONT VIEW

Fig. 13-27 Lugs

Shrink Rule

Patterns must be made larger than the desired size of the casting to compensate for the shrinkage which occurs when metal cools. A *shrink rule* is used by the patternmaker to overcome this difference. Different shrink rules are used for different materials.

Shrinkage allowances have been established for various metals:

- Cast iron and malleable iron — 1/8 inch per foot (10 mm per meter)
- Copper, aluminum, and bronze — 3/16 inch per foot (16 mm per meter)
- Steel — 1/4 inch per foot (21 mm per meter)
- Lead 5/16 inch per foot (26 mm per meter)

The patternmaker uses the shrink-rule measurement which correspond to the shrinkage allowance. In this way the pattern will be large enough to compensate for shrinkage, and the final casting will shrink to the desired original size.

UNIT 14

BASIC WELDING

WELDING

Welding is used in place of bolts. Screws, rivets and other types of fasteners. It is also used to fabricate parts that were formerly cast or forged. Welding is used to a considerable extent in the erection of frames for structural steel buildings, ships and other large structures.

Fusion Welding

In *fusion welding,* a welding rod is melted and combined with the metal parts that are to be fastened together. The parts will be permanently joined after cooling. The process can be done using torches or high electric power.

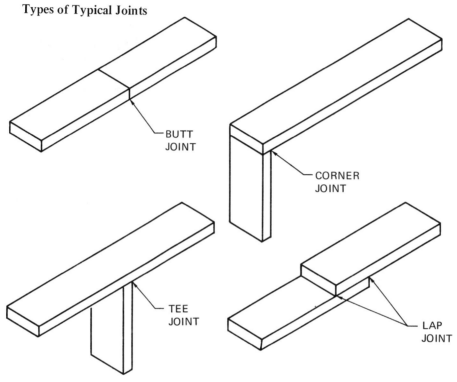

Types of Typical Joints

Fig. 14-1 Types of welded joints

Symbols

Figure 14-2 indicates the symbols for each type of weld. Figure 14-3 gives an example of each type of weld.

TYPE OF WELDS

① BACK OR BACKING WELD	② FILLET WELD	③ PLUG WELD	④ SQUARE WELD	⑤ V WELD	⑥ BEVEL WELD	⑦ U WELD	⑧ J WELD
⌒	◿	▭	‖	⋁	⋁	⋃	⋃

Fig. 14-2 Weld symbols

Figure 14-3 gives an example of each type of weld.

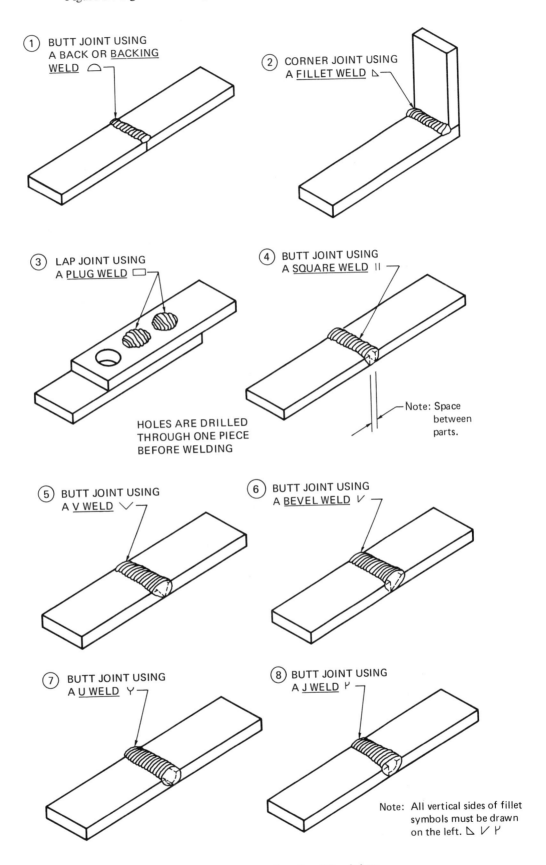

① BUTT JOINT USING A BACK OR BACKING WELD

② CORNER JOINT USING A FILLET WELD

③ LAP JOINT USING A PLUG WELD

HOLES ARE DRILLED THROUGH ONE PIECE BEFORE WELDING

④ BUTT JOINT USING A SQUARE WELD ‖

Note: Space between parts.

⑤ BUTT JOINT USING A V WELD

⑥ BUTT JOINT USING A BEVEL WELD

⑦ BUTT JOINT USING A U WELD

⑧ BUTT JOINT USING A J WELD

Note: All vertical sides of fillet symbols must be drawn on the left.

Fig. 14-3 Examples of welding joints

Placing Weld Symbols

Rule 1. When the weld symbol is placed below the reference line, figure 14-4(B), the weld appears on the same side as the arrowhead.

Rule 2. When the weld symbol is placed above the reference line (A), the weld appears on the opposite side of the arrowhead.

Rule 3. When the weld symbol is placed above and below the reference line (A and B), the weld appears on both sides of the arrowhead.

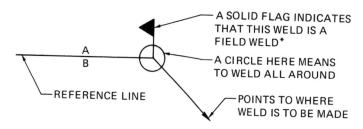

Fig. 14-4 Placing standard welding symbols

Figure 14-5 shows examples of rules 1, 2, and 3 using a fillet weld symbol.

Usually material over .125 inch (3) thick requires a *groove* (square, V, beveled, U, or J). Using the basic weld symbol, point the arrowhead toward the part that has the groove, figure 14-6.

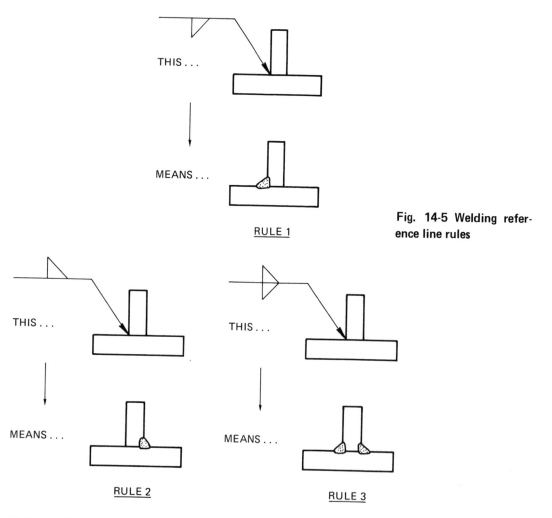

Fig. 14-5 Welding reference line rules

*A field weld symbol indicates that the weld must be made at the work site and not in the welding shop.

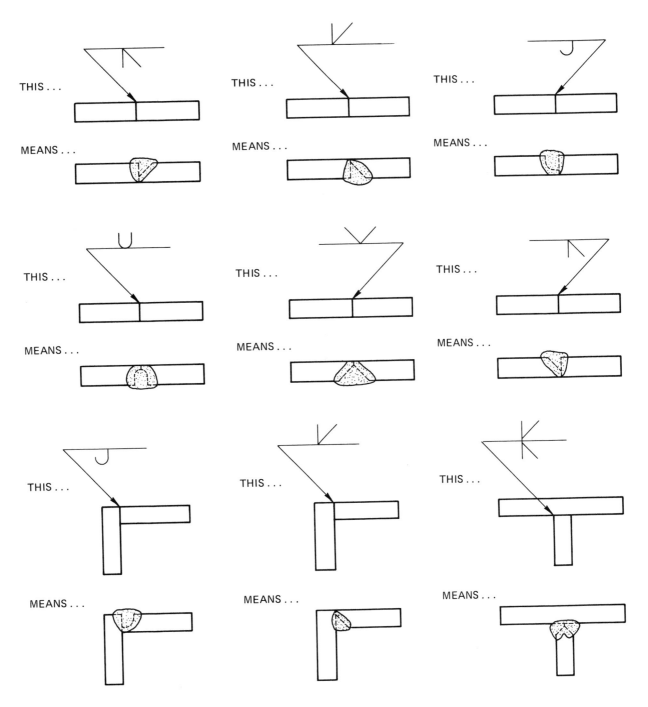

Groove welds shown require that the groove be machined to a specific size conforming to a specific delineation.

Fig. 14-6 Grooves in welding symbols

Reference Line Notations

There are various *notations* placed on or around the reference line. Figure 14-7 lists a few of the more widely used standard notations. Each tells the welder exactly how the drafter wants the part (s) welded. The .25″ X .375 (6 X 10) *size notation* means the weld is approximately .25″ X .375 and is welded the whole length of the part.

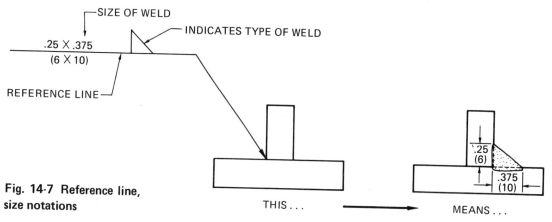

Fig. 14-7 Reference line, size notations

In figure 14-8, two more notations are added. The first means the length of each weld, and the second means the distance from center to center of each weld or pitch. *Pitch* refers to the distance, center to center, of each weld. The notation means that the weld is to be 2 inches long with a center-to-center distance of 4 inches. The notations would indicate millimetres if the metric system is used.

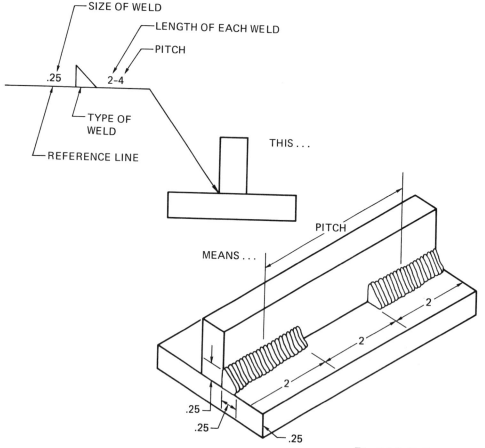

Fig. 14-8 Reference line, pitch notation

RESISTANCE WELDING

Resistance welding is the process of passing an electric current through a spot where the parts are to be joined. Symbols for resistance welding are shown in figure 14-9.

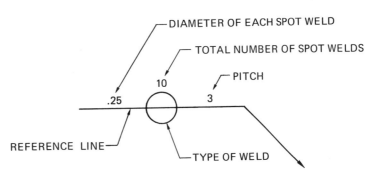

Fig. 14-9 Symbols for resistance welds

Study figure 14-10. Note that the symbols are similar to those used in fusion welding. As with fusion welding, notations are done on the side of the arrow where they appear.

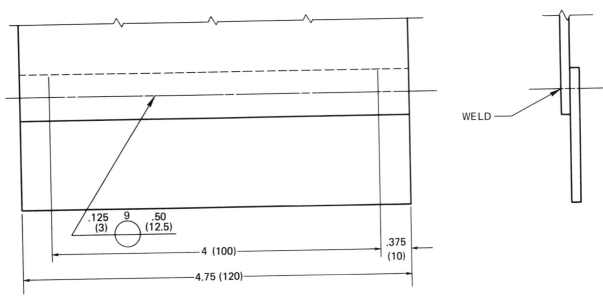

Fig. 14-10 Reference line for resistance welding

Spot Welding

Spot welding joins parts together with small circles or spots of heat. Figure 14-11 shows how a drafter would draw and dimension a drawing. Figure 14-12 shows how a welder would spot weld the project.

Fig. 14-11

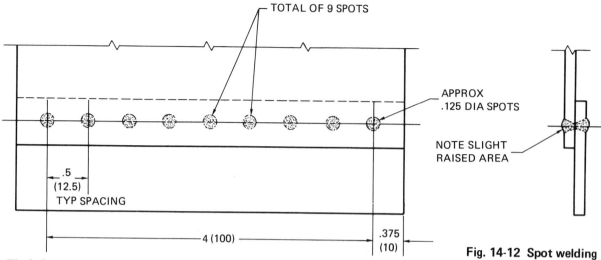

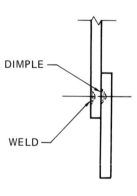

Fig. 14-12 Spot welding

Flush Symbol

A *flush symbol* is used to indicate that one or both surfaces must be ground smooth, figure 14-13.

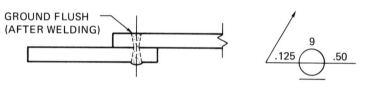

Fig. 14-13 Flush symbol added to welding symbol

Projection Weld

A *projection weld* is the same as a spot weld except one part has a *dimple* stamped into it at each spot where it is to be welded. This dimple allows more penetration and, as a result, is a better weld, figure 14-14.

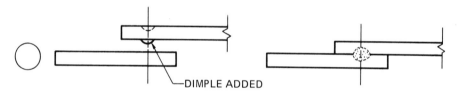

Fig. 14-14 Dimple symbol and application

Figure 14-15 shows how a drafter would draw and dimension a drawing. Note that the symbol is located below the reference line indicating that the dimple is on the part that is on the arrow side.

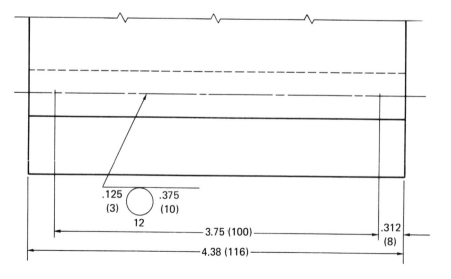

Fig. 14-15

Figure 14-16 shows how the welder would weld the project.

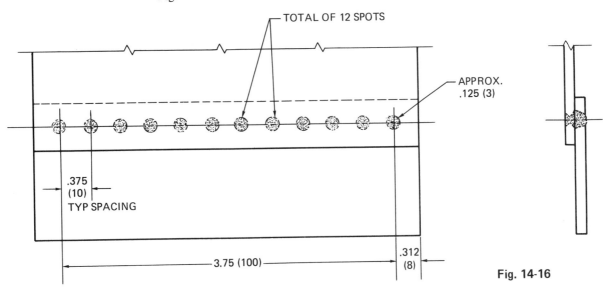

Fig. 14-16

Resistance Seam Weld

A *resistance seam weld* is like the spot weld process except the weld is continuous from start to finish. Figure 14-17 shows how the drafter would draw and dimension a drawing.

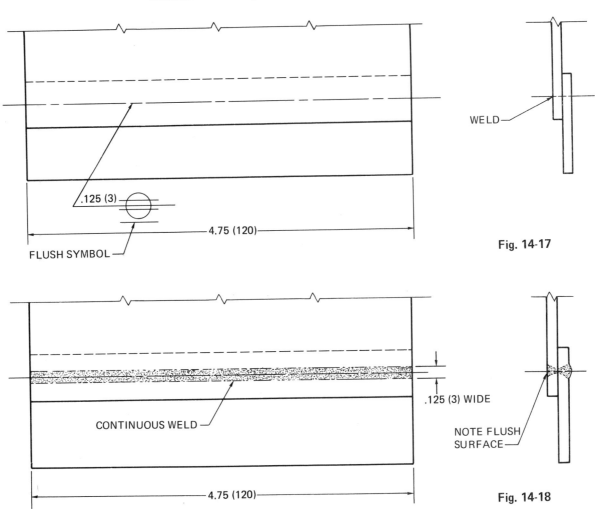

Fig. 14-17

Fig. 14-18

UNIT 15

FASTENERS

FASTENERS

Objects that are assembled must be held together with some type of fastener or by a fastening procedure. There are two major classifications of fasteners, permanent and temporary. *Permanent fasteners* are used when parts will not be disassembled. *Temporary fasteners* are used when the parts will be disassembled at some future time.

Permanent fastening methods include:

- Welding
- Brazing
- Stapling
- Nailing
- Gluing
- Riveting

Temporary fasteners include:

- Screws
- Bolts
- Keys
- Pins

There are many types and sizes of fasteners, each designed for a particular function. Welding procedures were discussed in Unit 2. Unit 3 provides detailed information on screw threads and the more common temporary fasteners. Although screw threads have other important uses, such as adjusting parts and transmitting power, only their use as a fastener will be discussed.

THREADS

Figure 15-1 illustrates a Unified National thread and labels the terms associated with all threads.

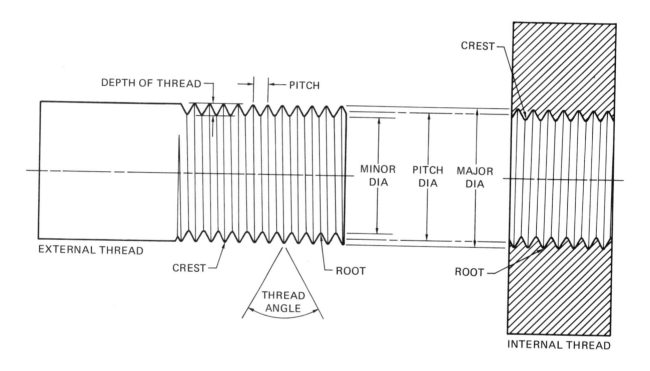

Fig. 15-1 Unified National thread

Common Thread Forms

Figure 15-2 illustrates the most common type of thread forms:

- Unified National
- Square — transmits power
- ACME — transmits power
- Buttress — transmits power in only one direction
- Worm — transmits power

Note that pitch (P) determines all other dimensions.

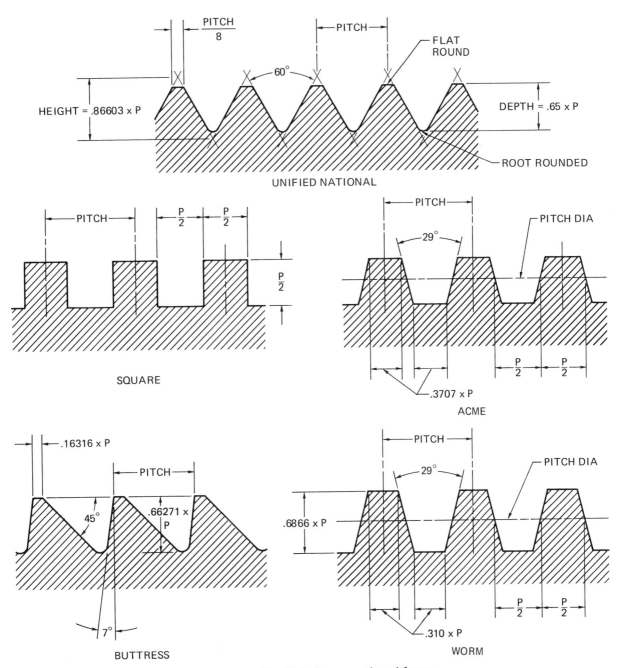

Fig. 15-2 Common thread forms

Classes of Fits

The fit of a thread is the amount of play between the screw and the nut when they are assembled together. There are three classes of fits for external threads and internal threads, figure 15-3.

CLASS OF FIT		CHARACTERISTICS
External Thread	Internal Thread	
1A	1B	Loosest fit. Used where easy assembly and disassembly is important.
2A	2B	Average fit. Used for ordinary fasteners.
3A	3B	Tight fit. Used when a snug fit is required for greater precision.

Fig. 15-3 Class of fit

TAP AND DIE

There are various methods to produce inside and outside threads. The simplest method uses thread cutting tools called *taps* and *dies.* A tap cuts internal threads, while a die cuts external threads, figures 15-4 and 15-5.

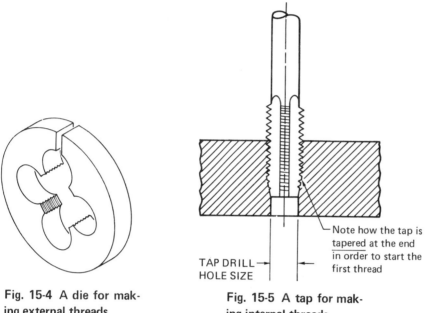

TAP DRILL → ← HOLE SIZE

└ Note how the tap is tapered at the end in order to start the first thread

Fig. 15-4 A die for making external threads

Fig. 15-5 A tap for making internal threads

Single and Multiple Threads

The *single thread* is standard. *Double* and *triple* threads are used when speed or travel distance is important, figure 15-6. The single thread, however, has more holding power.

A good example of double or triple threads is an inexpensive ball-point pen. Carefully take a ball-point pen apart and study the end of the external threads that hold the parts together. Note how fast the parts screw together. This is a characteristic of multiple threads.

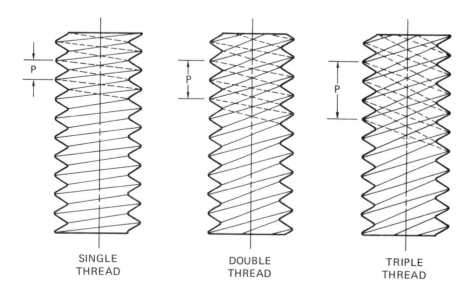

SINGLE THREAD   DOUBLE THREAD   TRIPLE THREAD

Fig. 15-6 Single and multiple threads

Thread Representation

Figure 15-7 shows a normal view of an *external thread* made with a threading die. Figure 15-8 is a representative drawing showing the information needed to manufacture the thread. The dash lines used to draw the minor diameter are not hidden edge lines but merely indicate the *depth of thread.*

Figure 15-9 shows a normal view of an *internal thread* made with a tap. Figure 15-10 shows the representative drawing made by the drafter. In the representative drawing, the minor diameter is made the same size as the diameter of the tap drill recommended. The major diameter in the drawing is the same diameter as the major diameter of the thread being represented.

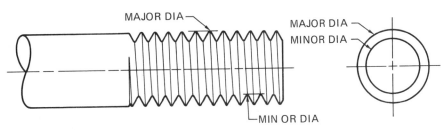

Fig. 15-7 External thread

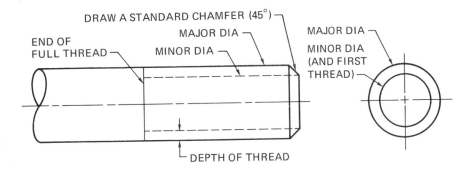

Fig. 15-8 External thread representation

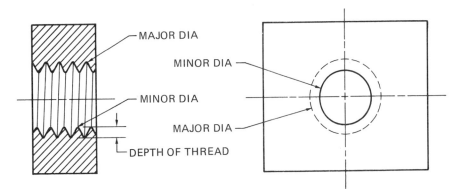

Fig. 15-9 Internal thread

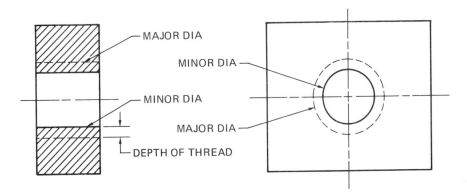

Fig. 15-10 Internal thread representation

Threaded Blind Hole

A *blind hole* is a hole drilled only part way through a piece of stock. Blind holes are often tapped to produce internal threads for screws. Figure 15-11 shows how blind holes are represented on drawings. The representative drawing is done to make sure the tap enters the stock deeply enough to assure the number of threads needed.

1. A hole is first drilled with a tap drill.
2. The tap is then turned into the tap hole.
3. The tap cuts threads as it enters the tap hole. Because the tap is tapered, it does not cut threads all the way to the bottom of the tap hole.

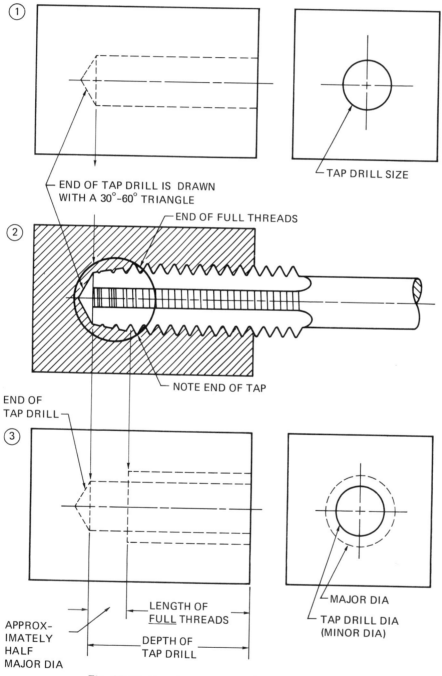

Fig. 15-11 How to represent a threaded blind hole

FASTENER CALLOUTS

Although all companies have not yet adopted the same standard for callouts, it is important that all workers within a company use the same method. The callout in figure 15-12 is the national standard callout.

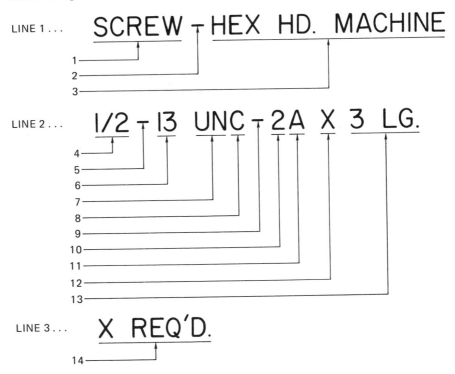

Fig. 15-12 Standard fastener callout

Line 1

1. Type of fastener
2. Dash
3. Description of fastener

Line 2

4. Nominal size
5. Dash
6. T.P.I. (threads per inch)
7. Thread series (Unified National)
8. Thread coarseness (F = fine, EF = extra fine, C = coarse)
9. Dash
10. Class of fit (1 = loose, 2 = average, 3 = tight)
11. External (A) *or* internal (B) thread
12. X = separation
13. Length (if required)

Line 3

14. Number of fasteners required

Examples of Fastener Callouts:

- NUT – HEX HD.
  5/8 – 18 UNF
  2 REQ'D
- WASHER – LOCK
  3/8
  1 REQ'D

- PIN – COTTER
  1/16 x 1 LG.
  10 REQ'D

At this time the United States, Canada, and the United Kingdon use the inch series of *Unified Thread Form.* This system designates the diameter and number of threads per inch along with a suffix indicating the thread series.

Example: 1/4 – 20 UNC

Threads are designated in the metric system using a similar approach, figure 15-13.

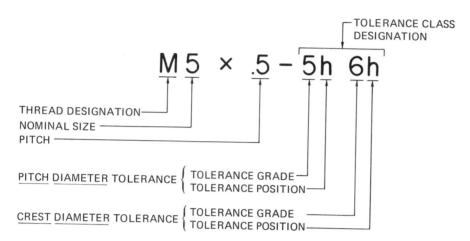

Fig. 15-13 Standard metric fastener callout

M = Shows the thread to be a metric thread
5 = Nominal size diameter in millimetres
X = Separation
.5 = Pitch of thread is 0.5 millimetres
– = Dash
5h = Tolerance class of pitch diameter
6h = Tolerance class of crest diameter

In the metric thread callout, a lower case h or g indicates *external threads,* while an upper H or G indicates *internal threads.*

In figure 15-13, the last two number and letter combinations refer to the *tolerance grades* and position of the *pitch diameter* and *crest diameter.* They are similar to the class of fit shown on line 10 in figure 15-12 dealing with threads in the inch system of callouts.

The diagram in figure 15-14 explains the symbols giving *class of fit* in the metric system.

Class of Fit	Tolerance Class	
	Bolts & Screws	Nuts
Fine	4h	5H
Medium	6g	6H
Coarse	8g	7H

Fig. 15-14 Class of fit — metric system

Metric thread callouts for fine threads use the prefix M, the diameter, and the pitch. When calling out coarse threads, only the prefix M and the diameter is used. The metric system uses coarse threads most frequently. Because the terminology of the inch system and the metric system is so different, refer to this text before adding any metric callouts to a drawing.

MEASURING THREADS

One method of measuring *threads per inch* (T.P.I.) is to place a standard scale on the crests of the threads and count the number of full threads within one inch of the scale, figure 15-15. If only part of an inch of stock is threaded, count the number of full threads in one-half inch and multiply by two to determine T.P.I.

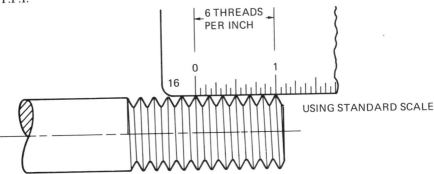

Fig. 15-15 Measuring threads with a standard scale

A simple, more accurate method of determining threads per inch is to use a *screw thread pitch gauge*, figures 15-16 and 15-17. By trial and error the various fingers or leaves of the gauge are placed over the threads until one is found that fits exactly into the threads. Threads per inch are then read directly on each leaf of the gauge.

Appendix B gives a dimension and size chart for the American National Standard Unified and American National Thread Series.

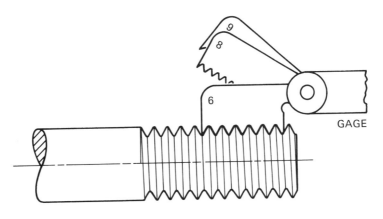

Fig. 15-16 Measuring threads using a screw thread pitch gauge

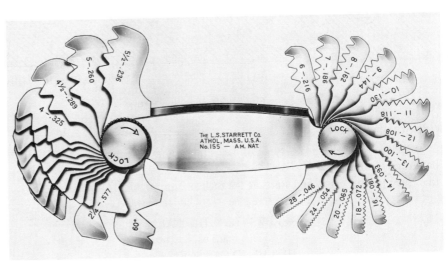

Fig. 15-17 Screw thread pitch gauge

RIGHT-HAND AND LEFT-HAND THREADS

Threads can be either right-handed or left-handed. In order to distinguish between a right-hand or left-hand thread, use this simple trick.

A right-hand thread winding tends to lean to the left, figure 15-18. If the thread leans to the left, the right-hand thumb points in the same direction. If the thread leans to the right, figure 15-19, the left-hand thumb leans in that direction indicating that it is a left-hand thread.

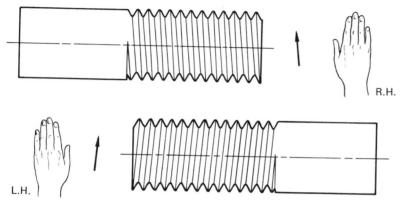

R.H.

Fig. 15-18 Right-hand thread

L.H.

Fig. 15-19 Left-hand thread

MACHINE SCREWS

Machine screws use a nut to fasten parts together. They also may be used in a tapped hole, figure 15-20. The five standard head types are flat, round, oval, fillister, and pan. They range in numbered sizes from #0 to #12 and in fractional sizes from 1/4 inch to 3/4 inch in length.

A typical callout for a machine screw is:

SCREW – RD. HD. MACH.
This is read as:
ROUND HEAD MACHINE SCREW

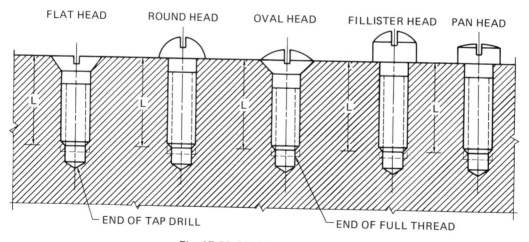

FLAT HEAD   ROUND HEAD   OVAL HEAD   FILLISTER HEAD   PAN HEAD

END OF TAP DRILL   END OF FULL THREAD

Fig. 15-20 Machine screws

CAP SCREWS

There are five standard types of heads under the classification of *cap screws:* flat, round, fillister, hex, and socket head, figure 15-21. The cap screw is larger than the machine screw and starts at 1/4 inch size through 1 1/4 inch size.

A typical callout for a cap screw is:

SCREW – HEX. HD. CAP
This reads as:
HEX HEAD CAP SCREW

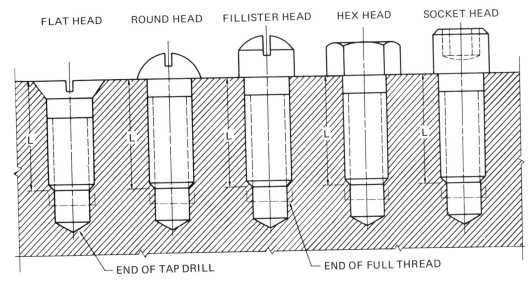

Fig. 15-21 Cap screws

FLAT-HEAD SCREW

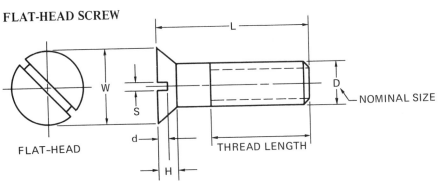

Fig. 15-22 Flat-head screw representation

Type	Nom. Size (Inch)	— D —		— W —	— H —	— S —	— d —
		Inch	mm	Inch	Inch	Inch	Inch
Mach.	0	.060	1.524	.119	.035	.023	.015
	1	.073	1.854	.146	.043	.026	.019
	2	.086	2.108	.172	.051	.031	.023
	3	.099	2.514	.199	.059	.035	.027
	4	.112	2.845	.225	.067	.039	.030
	5	.125	3.175	.252	.075	.043	.034
	6	.138	3.505	.279	.083	.048	.033
	8	.164	4.166	.332	.100	.054	.045
	10	.190	4.826	.385	.116	.060	.053
	12	.216	5.486	.438	.132	.067	.060
Cap	1/4	.250	6.350	.507	.153	.075	.070
	5/16	.313	7.950	.635	.191	.084	.088
	3/8	.375	9.525	.762	.230	.094	.106
	7/16	.438	11.125	.812	.223	.094	.103
	1/2	.500	12.700	.875	.223	.106	.103
	9/16	.563	14.300	1.000	.260	.118	.120
	5/8	.625	15.875	1.125	.298	.133	.137
	3/4	.750	19.050	1.375	.372	.149	.171

Fig. 15-23 Flat-head screw size chart

All sizes are maximum limit

ROUND-HEAD SCREWS

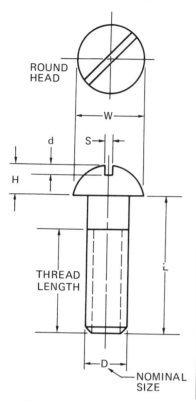

Fig. 15-24 Round-head
screw representation

Type	Nom. Size (Inch)	– D –		– W –	– H –	– S –	– d –
		Inch	mm	Inch	Inch	Inch	Inch
Mach.	0	.060	1.524	.113	.053	.023	.039
	1	.073	1.854	.138	.061	.026	.044
	2	.086	2.184	.162	.069	.031	.048
	3	.099	2.514	.187	.078	.035	.053
	4	.112	2.845	.211	.086	.039	.058
	5	.125	3.175	.236	.095	.043	.063
	6	.138	3.505	.260	.103	.048	.068
	8	.164	4.166	.309	.120	.054	.077
	10	.190	4.826	.359	.137	.060	.087
	12	.216	5.486	.408	.153	.067	.096
Cap	1/4	.250	6.350	.472	.175	.075	.109
	5/16	.313	7.950	.590	.216	.084	.132
	3/8	.375	9.525	.708	.256	.094	.155
	7/16	.438	11.125	.750	.328	.094	.196
	1/2	.500	12.700	.813	.355	.106	.211
	9/16	.563	14.300	.938	.410	.118	.242
	5/8	.625	15.875	1.000	.438	.133	.258
	3/4	.750	19.050	1.250	.547	.149	.320

All sizes are maximum limit

Fig. 15-25 Round-head screw size chart

FILLISTER-HEAD SCREWS

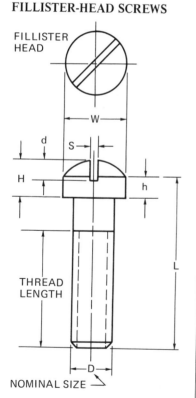

Fig. 15-26 Fillister-head
screw representation

Type	Nom. Size (Inch)	– D –		– W –	– H –	– h –	– S –	– d –
		Inch	mm	Inch	Inch	Inch	Inch	Inch
Mach.	0	.060	1.524	.096	.059	.045	.023	.025
	1	.073	1.854	.118	.071	.053	.026	.031
	2	.086	2.184	.140	.083	.062	.031	.037
	3	.099	2.514	.161	.095	.070	.035	.043
	4	.112	2.845	.183	.107	.079	.039	.048
	5	.125	3.175	.205	.120	.088	.043	.054
	6	.138	3.505	.226	.132	.096	.048	.060
	8	.164	4.166	.270	.156	.113	.054	.071
	10	.190	4.826	.313	.180	.130	.060	.083
	12	.216	5.486	.357	.205	.148	.067	.094
Cap	1/4	.250	6.350	.414	.237	.170	.075	.109
	5/16	.313	7.950	.518	.295	.211	.084	.137
	3/8	.375	9.525	.622	.355	.253	.094	.164
	7/16	.438	11.125	.625	.368	.265	.094	.170
	1/2	.500	12.700	.750	.412	.297	.106	.190
	9/16	.563	14.300	.812	.466	.336	.118	.214
	5/8	.625	15.875	.875	.521	.375	.133	.240
	3/4	.750	19.050	1.000	.612	.441	.149	.281

All sizes are maximum limit

Fig. 15-27 Fillister-head screw size chart

OVAL-HEAD SCREWS

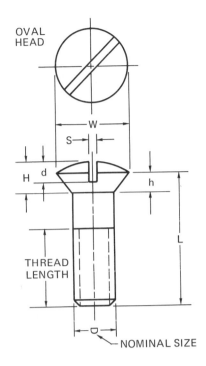

Fig. 15-28 Oval-head screw representation

Type	Nom. Size (Inch)	— D —		— W —	— H —	— h —	— S —	— d —
		Inch	mm	Inch	Inch	Inch	Inch	Inch
Mach.	0	.060	1.524	.119	.056	.035	.023	.030
	1	.073	1.854	.146	.068	.043	.026	.038
	2	.086	2.184	.172	.080	.051	.031	.045
	3	.099	2.514	.199	.092	.059	.035	.052
	4	.112	2.845	.225	.104	.067	.039	.059
	5	.125	3.175	.252	.116	.075	.043	.067
	6	.138	3.505	.279	.128	.083	.048	.074
	8	.164	4.166	.332	.152	.100	.054	.088
	10	.190	4.826	.385	.176	.116	.060	.103
	12	.216	5.486	.438	.200	.132	.067	.117
Cap	1/4	.250	6.350	.507	.232	.153	.075	.136
	5/16	.313	7.950	.635	.290	.191	.084	.171
	3/8	.375	9.525	.762	.347	.230	.094	.206
	7/16	.438	11.125	.812	.345	.223	.094	.210
	1/2	.500	12.700	.875	.354	.223	.106	.216
	9/16	.563	14.300	1.000	.410	.260	.118	.250
	5/8	.625	15.875	1.125	.467	.298	.133	.285
	3/4	.750	19.050	1.375	.578	.372	.149	.353

All sizes are maximum limit

Fig. 15-29 Oval-head screw size chart

HEX-HEAD SCREWS AND BOLTS

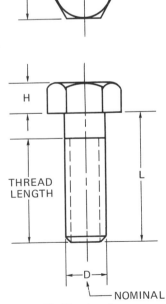

Fig. 15-30 Hex-head representation

Type	Nom. Size (Inch)	— D —		— W —		— H —	
		Inch	mm	Inch	mm	Inch	mm
Cap Screws	1/4	.260	6.60	.438	11.11	.163	4.14
	5/16	.324	8.23	.500	12.70	.211	5.36
	3/8	.388	9.86	.563	14.28	.243	6.17
	7/16	.452	11.48	.625	15.87	.291	7.39
	1/2	.515	13.08	.750	19.05	.323	8.20
	9/16	.577	14.66	.813	20.64	.371	9.42
	5/8	.642	16.31	.938	23.81	.403	10.24
	3/4	.768	19.51	1.125	28.57	.483	12.26
	7/8	.895	22.73	1.313	33.34	.563	14.30
	1	1.022	25.95	1.500	38.10	.627	15.93
	1 1/8	1.149	29.18	1.688	42.86	.718	18.23
	1 1/4	1.277	32.43	1.875	47.63	.813	20.65
Bolts	1 3/8	1.404	35.66	2.063	52.38	.940	23.88
	1 1/2	1.531	38.88	2.250	57.15	1.036	26.31
	1 3/4	1.785	45.34	2.625	66.68	1.196	30.37
	2	2.039	51.79	3.000	76.20	1.388	35.25
	2 1/4	2.305	58.54	3.375	85.72	1.548	39.32
	2 1/2	2.559	64.99	3.750	95.25	1.708	43.38
	2 3/4	2.827	71.80	4.125	104.78	1.869	47.47
	3	3.081	78.25	4.500	114.30	2.060	52.32

Fig. 15-31 Hex-head screw and bolt size chart

SQUARE-HEAD SCREWS AND BOLTS

SQUARE HEAD

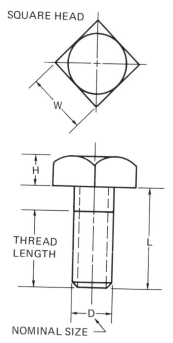

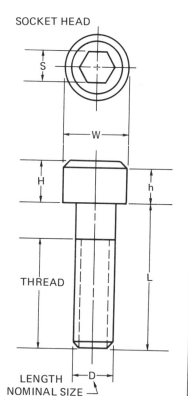

THREAD LENGTH

NOMINAL SIZE

Fig. 15-32 Square-head representation

Type	Nom. Size (Inch)	– D –		– W –		– H –	
		Inch	mm	Inch	mm	Inch	mm
Cap Screws	1/4	.260	6.60	.375	9.52	.188	4.77
	5/16	.324	8.23	.500	12.70	.220	5.58
	3/8	.388	9.86	.563	14.28	.268	6.81
	7/16	.452	11.48	.625	15.87	.316	8.03
	1/2	.515	13.08	.750	19.05	.348	8.84
	5/8	.642	16.31	.938	23.81	.444	11.28
	3/4	.768	19.51	1.125	28.57	.524	13.31
	7/8	.895	22.73	1.313	33.34	.620	15.75
	1	1.022	25.95	1.500	38.10	.684	17.38
	1 1/8	1.149	29.18	1.688	42.86	.780	19.81
	1 1/4	1.277	32.43	1.875	47.63	.876	22.22
Bolts	1 3/8	1.404	35.66	2.063	52.38	.940	23.88
	1 1/2	1.531	38.88	2.250	57.15	1.036	26.31
	1 5/8	1.658	42.11	2.438	61.91	1.132	28.75

Fig. 15-33 Square-head screw and bolt size chart

SOCKET-HEAD SCREWS

SOCKET HEAD

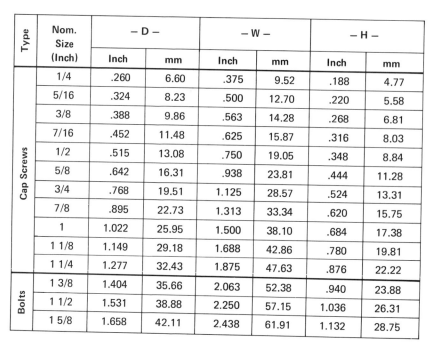

THREAD

LENGTH
NOMINAL SIZE

Fig. 15-34 Socket-head screw representation

Type	Nom. Size (Inch)	– D –		– W –	– H –	– h –	– S –
		Inch	mm	Inch	Inch	Inch	Inch
Mach.	0	.060	1.524	.096	.060	.055	.051
	1	.073	1.854	.118	.073	.067	.051
	2	.086	2.184	.140	.086	.079	.063
	3	.099	2.514	.161	.099	.091	.079
	4	.112	2.845	.183	.112	.103	.079
	5	.125	3.175	.205	.125	.115	.095
	6	.138	3.505	.226	.138	.127	.095
	8	.164	4.166	.270	.164	.150	.127
	10	.190	4.826	.313	.190	.174	.158
	12	.216	5.486	.344	.216	.198	.158
Cap	1/4	.250	6.350	.375	.250	.229	.189
	5/16	.313	7.950	.438	.313	.286	.220
	3/8	.375	9.525	.563	.375	.344	.315
	7/16	.438	11.125	.625	.438	.401	.315
	1/2	.500	12.700	.750	.500	.458	.378
	9/16	.563	14.300	.813	.563	.516	.378
	5/8	.625	15.875	.875	.625	.573	.503
	3/4	.750	19.050	1	.750	.688	.565

All sizes maximum limit

Fig. 15-35 Socket-head screw size chart

SETSCREWS

Setscrews are used to prevent rotary motion between mating parts, such as a pulley and shaft. They come in slotted, hex-socket, and square heads, figure 15-36. Note how all sizes are derived from the diameter (D).

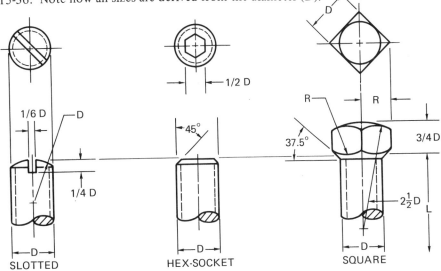

Fig. 15-36 Setscrew head types

SLOTTED   HEX-SOCKET   SQUARE

Points

Setscrew head styles come with a flat, cup, oval, cone, half-dog, or full-dog point, figure 15-37.

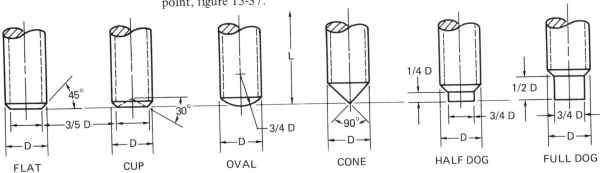

FLAT   CUP   OVAL   CONE   HALF DOG   FULL DOG

Fig. 15-37 Setscrew point types

Type	Nom. Size (Inch)	– D –		– UNC – (Coarse)	– UNF – (Fine)
		Inch	M.M.		
Mach.	0	.060	1.524	–	80
	1	.073	1.854	64	72
	2	.086	2.184	56	64
	3	.099	2.514	48	56
	4	.112	2.845	40	48
	5	.125	3.175	40	44
	6	.138	3.505	32	40
	8	.164	4.166	32	36
	10	.190	4.826	24	32
	12	.216	5.486	24	28
Cap	1/4	.250	6.350	20	28
	5/16	.313	7.950	18	24
	3/8	.375	9.525	16	24
	7/16	.438	11.125	14	20
	1/2	.500	12.700	13	20
	9/16	.563	14.300	12	18
	5/8	.625	15.875	11	18

Fig. 15-38 Setscrew size chart

HEXAGONAL AND SQUARE NUTS

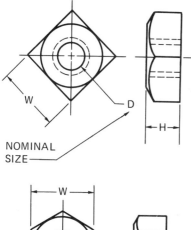

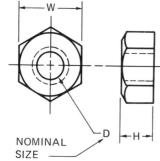

NOMINAL SIZE

Fig. 15-39 Nut representation

Nom. Size (Inch)	– D – Inch	– D – mm	– W – Inch	– W – mm	– H – Inch	– H – mm
1/4	.260	6.60	.438	11.11	.235	5.96
5/16	.324	8.23	.563	14.28	.283	7.18
3/8	.388	9.86	.625	15.87	.346	8.78
7/16	.452	11.48	.750	19.05	.394	10.00
1/2	.515	13.08	.813	20.64	.458	11.63
9/16	.577	14.66	.875	22.23	.521	13.23
5/8	.642	16.31	1	25.40	.569	14.45
3/4	.768	19.51	1.125	28.57	.680	17.27
7/8	.895	22.73	1.313	33.34	.792	20.12
1	1.022	25.95	1.500	38.10	.903	22.94
1 1/8	1.149	29.18	1.688	42.86	1.030	26.16
1 1/4	1.277	32.43	1.875	47.63	1.126	28.60
1 3/8	1.404	35.66	2.063	52.38	1.237	31.42
1 1/2	1.531	38.88	2.250	57.15	1.348	34.24

Fig. 15-40 Nut size chart

PLAIN WASHERS

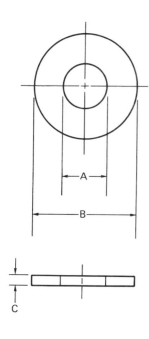

Fig. 15-41 Washer representation

Tol.	– A – Inch	– A – mm	Tol.	– B – Inch	– B – mm	Nom.	– C – Inch	– C – mm
+OR −.005 (.127 M.M.)	3/32	2.38		1/4	6.35		.020	.51
	1/8	3.18		5/16	7.94		.032	.81
	5/32	3.97		3/8	9.53		.049	1.24
	3/16	4.76		7/16	11.13		.049	1.24
	7/32	5.56		7/16	11.13		.049	1.24
+OR −.010 (.254 M.M.)	1/4	6.35	+OR −.010 (.254 M.M.)	1/2	12.70	NOMINAL SIZE	.049	1.24
	9/32	7.14		5/8	15.88		.065	1.65
	5/16	7.94		7/8	22.23		.065	1.65
	11/32	8.73		11/16	17.47		.065	1.65
	3/8	9.53		3/4	19.05		.065	1.65
	13/32	10.32		13/16	20.64		.065	1.65
	7/16	11.11		1	25.40		.083	2.11
	1/2	12.70		1 1/4	31.75		.083	2.11
	17/32	13.50		1 1/16	26.99		.095	2.41
	9/16	14.28		1 3/8	34.93		.109	2.77
	5/8	15.88		1 3/8	34.93		.109	2.77
	11/16	17.46		1 1/2	38.10		.134	3.40
	13/16	20.64		1 1/2	38.10		.134	3.40
	15/16	23.81		2	50.80		.165	4.19
	1 1/16	26.98		2 1/4	57.15		.165	4.19

Fig. 15-42 Washer size chart

LOCK WASHERS

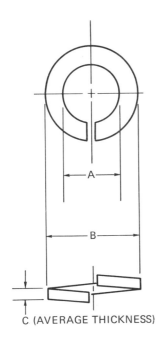

Fig. 15-43 Lock washer representation

– A –		– B –		– C –	
Inch	mm	Inch	mm	Inch	mm
.088	2.23	.175	4.44	.020	.51
.102	2.59	.198	5.02	.025	.64
.115	2.92	.212	5.38	.025	.64
.128	3.25	.239	6.07	.031	.79
.141	3.58	.253	6.43	.031	.79
.168	4.26	.296	7.52	.040	1.02
.194	4.92	.337	8.56	.047	1.02
.221	5.61	.380	9.65	.056	1.42
.255	6.47	.493	12.52	.062	1.57
.319	8.10	.591	15.01	.078	1.98
.382	9.70	.688	17.48	.094	2.38
.446	11.32	.784	19.91	.109	2.76
.509	12.92	.879	22.33	.125	3.18
.573	14.55	.979	24.87	.141	3.58
.636	16.15	1.086	27.58	.156	3.96
.700	17.78	1.184	30.07	.172	4.37
.763	14.38	1.279	32.48	.188	4.78
.827	21.01	1.377	34.97	.203	5.15
.890	22.60	1.474	37.44	.219	5.56
1.017	25.83	1.672	42.47	.250	6.35

Fig. 15-44 Lock washer size chart

COTTER PIN

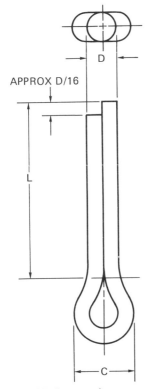

Fig. 15-45 Cotter pin representation

Nom. Size (Inch)	– D –		– d –		Hole Size	
	Inch	m.m.	Inch	m.m.	Inch	m.m.
1/32	.031	.787	.06	1.52	.047	1.193
3/64	.047	1.193	.09	2.28	.062	1.574
1/16	.062	1.574	.12	3.04	.078	2.006
5/64	.078	1.981	.16	4.06	.094	2.387
3/32	.094	2.387	.19	4.82	.109	2.768
7/64	.109	2.768	.22	5.58	.125	3.175
1/8	.125	3.175	.25	6.35	.141	3.581
9/64	.141	3.581	.28	7.11	.156	3.962
5/32	.156	3.962	.31	7.87	.172	4.368
3/16	.188	4.775	.38	9.65	.203	5.156
7/32	.219	5.562	.44	11.17	.234	5.943
1/4	.250	6.350	.50	12.70	.266	6.756
5/16	.312	7.925	.62	15.74	.312	7.925
3/8	.375	9.525	.75	19.05	.375	9.525
7/16	.438	11.125	.88	22.55	.438	11.125
1/2	.500	12.700	1.00	25.40	.500	12.700
5/8	.625	15.875	1.25	31.750	.625	15.875
3/4	.750	19.050	1.50	38.100	.750	19.050

Fig. 15-46 Cotter pin size chart

KEYS AND KEYSEATS

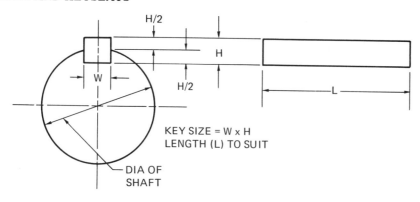

KEY SIZE = W x H
LENGTH (L) TO SUIT

DIA OF
SHAFT

Fig. 15-47 Key representation

Shaft Nom. Size – DIA. –		Square (W = H)	Type	Square Key		Tolerance
From	To & Incl.			From	To & Incl.	
5/16 (8)	7/16 (11)	3/32 (2.38)	Bar Stock	—	3/4 (19.05)	+.000 – .002 (+.0000 – .0254)
7/16 (11)	9/16 (14)	1/8 (3.175)		3/4 (19.05)	1 1/2 (38.1)	+.000 – .003 (+.0000 – .0762)
9/16 (14)	7/8 (22)	3/16 (4.76)		1 1/2 (38.1)	2 1/2 (63.5)	+.000 – .004 (+.0000 – .1016)
7/8 (22)	1 1/4 (32)	1/4 (6.35)		2 1/2 (63.5)	3 1/2 (88.9)	+.000 – .006 (+.0000 – .1524)
1 1/4 (32)	1 3/8 (35)	5/16 (7.94)	Keystock	—	1 1/4 (31.75)	+.001 – .000 (+.0254 – .0000)
1 3/8 (35)	1 3/4 (44)	3/8 (9.53)		1 1/4 (31.75)	3 (76.2)	+.002 – .000 (+.0508 – .0000)
1 3/4 (44)	2 1/4 (57)	1/2 (12.7)		3 (76.2)	3 1/2 (88.9)	+.003 – .000 (+.0762 – .0000)
2 1/4 (57)	2 3/4 (70)	5/8 (15.88)				
2 3/4 (70)	3 1/4 (82)	3/4 (19.05)				
3 1/4 (82)	3 3/4 (95)	7/8 (22.23)				

(Figures in parenthesis = mm)

Fig. 15-48 Key size chart

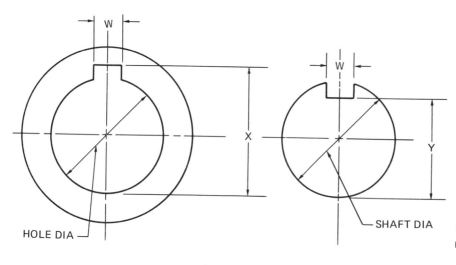

HOLE DIA

SHAFT DIA

Fig. 15-49 Keyway representation

Nom. Size	– DIA. – (Shaft)		'X' (Collar)		'Y' (Shaft)	
(Inch)	Inch	mm	Inch	mm	Inch	mm
1/2	.500	12.700	.560	14.224	.430	10.922
9/16	.562	14.290	.623	15.824	.493	12.522
5/8	.625	15.875	.709	18.008	.517	13.132
11/16	.688	17.470	.773	18.618	.581	14.757
3/4	.750	19.050	.837	21.259	.644	16.357
13/16	.812	20.640	.900	22.860	.708	17.983
7/8	.875	22.225	.964	24.485	.771	19.583
15/16	.938	23.820	1.051	26.695	.791	20.091
1	1.000	25.400	1.114	28.295	.859	21.818
1 1/16	1.062	26.985	1.178	29.921	.923	23.444
1 1/8	1.125	28.575	1.241	31.521	.986	25.044
1 3/16	1.188	30.165	1.304	33.121	1.049	26.644
1 1/4	1.250	31.750	1.367	34.722	1.112	28.244
1 5/16	1.312	33.340	1.455	36.957	1.137	28.879
1 3/8	1.375	34.923	1.518	38.557	1.201	30.505

Fig. 15-50 Keyway size chart

DRAWING FASTENERS

When drawing various fasteners, the drafter must use standard drawing practices.

In drawing a hexagonal nut or bolt head, always draw both the front view and the right-size view, figure 15-51, even though the right-side view is actually incorrect as shown. If drawn correctly, the hexagonal head actually looks like a square head.

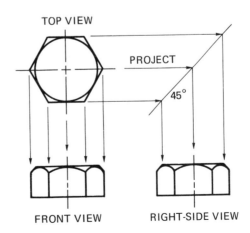

Fig. 15-51

Never draw a nut and bolt in section, figure 15-52, even if the cutting-plane line goes through the nut and bolt.

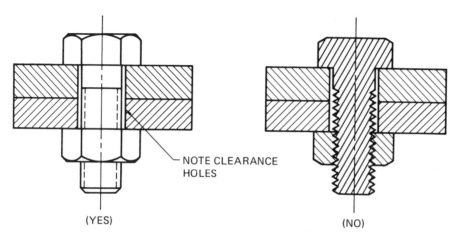

NOTE CLEARANCE HOLES

(YES)   (NO)

Fig. 15-52

A stud has two classes of fit. The section that screws into the part should have a tighter fit than the section at the nut end, figure 15-53. This is so the nut can be disassembled without removing the stud from the part. Sometimes fine threads are used at the stud end.

STUD END (CLASS 3 FIT)   NUT END (CLASS 2 FIT)

Fig. 15-53

When drawing various screw heads, draw the top view with the groove at a 45-degree angle as shown at B, figure 15-54, and the front view as if it were in position A. This is the standard method.

B

A

Fig. 15-54

The *standard thread depth* allowed in order to have the required strength is determined by the kind of material used. In steel, the depth *could* be thread diameter (minimum), but, in plastic or any other soft material, two times the thread diameter (minimum) must be used, figure 15-55. If possible, try to design it with more than minimum requirements, especially where safety is involved.

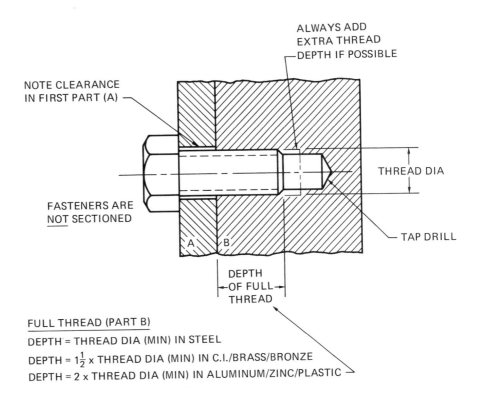

FULL THREAD (PART B)

DEPTH = THREAD DIA (MIN) IN STEEL

DEPTH = $1\frac{1}{2}$ x THREAD DIA (MIN) IN C.I./BRASS/BRONZE

DEPTH = 2 x THREAD DIA (MIN) IN ALUMINUM/ZINC/PLASTIC

Fig. 15-55

Any standard bolt or stud is usually designed around the thread lengths given in figure 15-56. The actual length is determined by the thread series (coarse, fine, or extra fine).

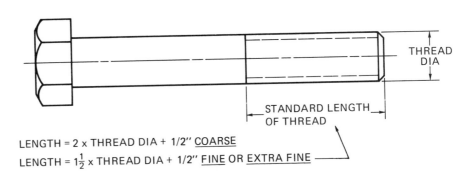

LENGTH = 2 x THREAD DIA + 1/2" <u>COARSE</u>

LENGTH = $1\frac{1}{2}$ x THREAD DIA + 1/2" <u>FINE</u> OR <u>EXTRA FINE</u>

Fig. 15-56

Thread Undercut or Relief

A simple spacer with threads looks like figure 15-57.

If an arm were attached and held in place with a nut, there probably will be a space between the shoulder and the arm, figure 15-58.

In order to eliminate this possibility, add an *undercut* or *thread relief,* figure 15-59.

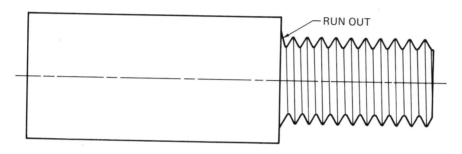

Fig. 15-57

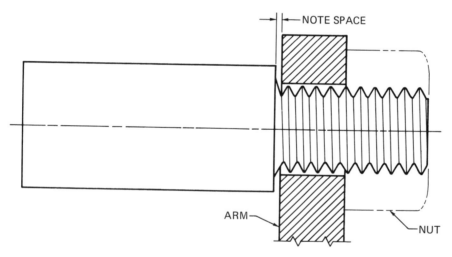

Fig. 15-58

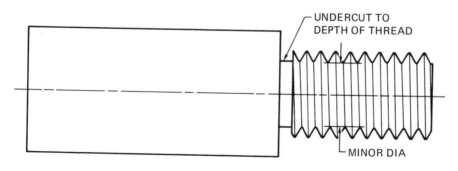

Fig. 15-59

Now the arm will fit against the shoulder tightly, figure 15-60. The thread relief is usually designed to the depth of the threads and called out as:

"XX WIDE X.XX DEEP THREAD RELIEF"

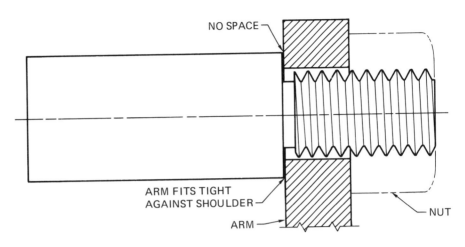

Fig. 15-60

Drawing a Fastener

The fastener callout and corresponding size charts provide the information needed to lay out a fastener. The drawing procedure given here uses a 1-UNC X 3 1/2 LG hexagonal bolt and nut as an example. For this fastener, refer to the chart for hex-head bolts on page 225 and the chart for hexagonal nuts on page 228.

Step 1. Lightly lay out the head diameter, length, and overall size of the nut.

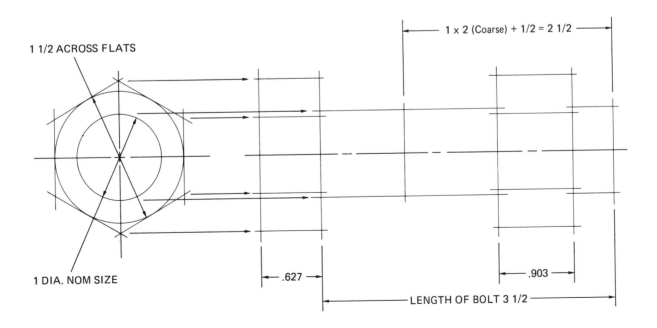

Step 2. Locate all radii, arcs, and chamfers as shown.

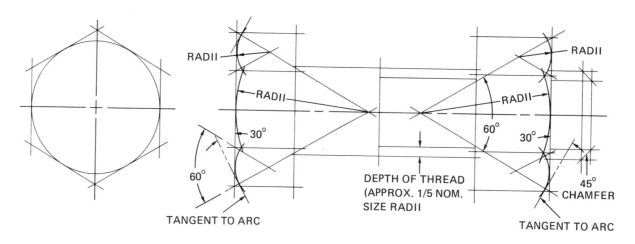

Step 3. Check all dimensions. Complete the drawing, using correct line weight.

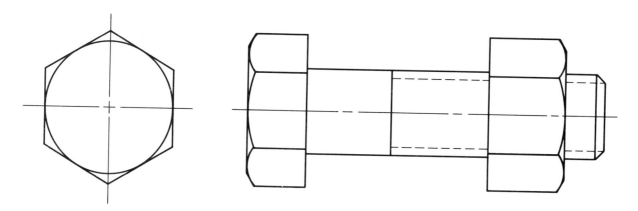

Review the charts on pages 223 through 231 once again. Note how each fastener is represented.

UNIT 16

PRECISION
MEASUREMENT

MEASURING

Unit 16 illustrates how to make precise measurements using both the inch (English system) and the millimetre (metric system). The metric system uses the *metre* (m) as its basic dimension. It is 3.281 feet long or about 3 3/8 inches longer than a yardstick. Its multiples, or parts, are expressed by adding prefixes. These prefixes represent equal steps of 1000 parts. The prefix for a thousand (1000) is *kilo;* the prefix for a thousandth (1/1000) is *milli.* One thousand metres (1000m), therefore, equals one kilometre (1.0 km). One thousandth of a metre (1/1000 m) equals one millimetre (1.0 mm). Comparing metric to English then:

One millimetre (1.0 mm) = 1/1000 metre - .03937 inch
One thousand millimetres (1000 mm) = 1.0 metre (1.0 m) = 3.281 feet
One thousand metres (1000 m) = 1.0 kilometre (1.0 km) = 3281.0 feet

POCKET STEEL RULER

The pocket steel ruler is the easiest of all measuring tools to use. The *inch scale*, figure 16-1, is six inches long and is graduated in 10ths and 100ths of an inch on one side and 32nds and 64ths on the other side.

The *metric scale* is 150 millimetres long (approximately six inches) and is graduated in millimetres and half millimetres on one side, figure 16-2. Sometimes metric pocket steel rulers are graduated in 64ths of an inch on the other side.

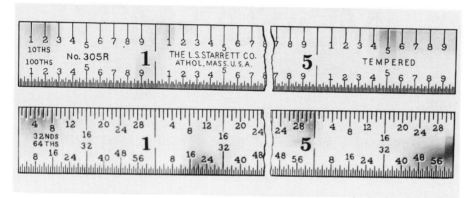

Fig. 16-1 Inch pocket ruler

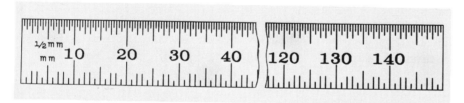

Fig. 16-2 Metric pocket ruler

CALIPERS AND DIVIDERS

Figures 16-3 and 16-4 show two types of calipers used in a machine shop. They are spring-loaded with a nut to lock measurements into position. The *inside caliper,* figure 16-3, takes internal measurements, such as hole diameters and groove widths. The *outside caliper,* figure 16-4 takes external measurements, such as rod diameters and stock thickness measurements. Figure 16-5 shows a pair of dividers, a tool used to measure and transfer distances.

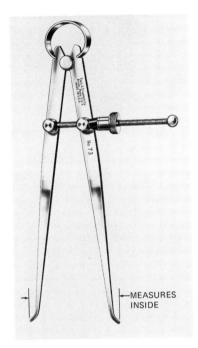

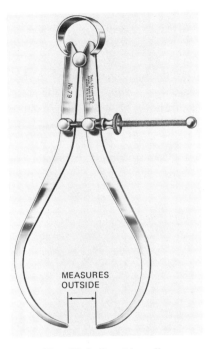

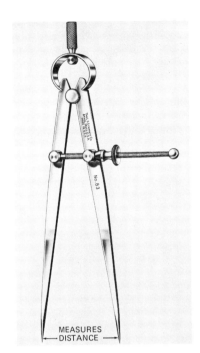

Fig. 16-3 Inside caliper   **Fig. 16-4 Outside caliper**   **Fig. 16-5 Dividers**

The calipers shown in figures 16-6 and 16-7 are similar to the calipers in figures 16-3 and 16-4 except that these are not spring-loaded. They are more sensitive and care must be taken not to alter the measurement once it is made with the caliper.

Fig. 16-6 Outside caliper   **Fig. 16-7 Inside caliper**

MICROMETER

The *micrometer* is a highly accurate screw which rotates inside a fixed nut, figure 16-8. As this screw is turned, it opens or closes a distance between the *measuring faces* of the anvil and spindle.

Place whatever is to be measured by the micrometer between the *anvil* and the *spindle*. Rotate the spindle by means of the *thimble* until the anvil and spindle come in contact with what is to be measured. The size is then determined by reading off the figures located on the *sleeve* and *thimble*.

Micrometers come in both English and metric graduations. They are manufactured with an English size range of 1 inch through 60 inches and a metric size range of 25 millimetres to 1500 millimetres. The micrometer is a very sensitive device and must be treated with extreme care.

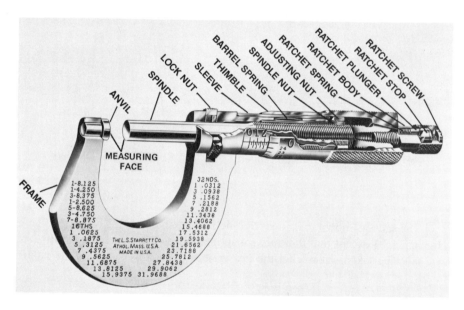

Fig. 16-8 Micrometer

Micrometers come in various shapes and sizes, figures 16-9 through 16-13.

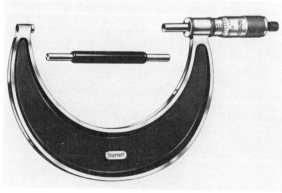

1″ TO 6″ MICROMETER

Fig. 16-9

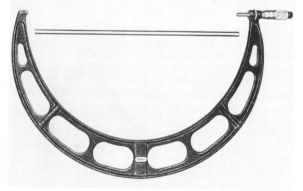

7″ TO 60″ MICROMETER

Fig. 16-10

Measurements of inside surfaces are made with a *telescoping gauge,* figure 16-12. The inside size is "locked," then the distance between ends is measured with a standard micrometer.

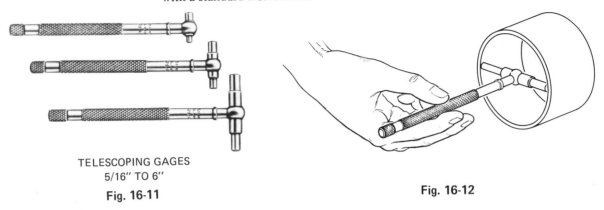

TELESCOPING GAGES
5/16" TO 6"

Fig. 16-11

Fig. 16-12

The depth of holes, slots, and various projections are measured with a *micrometer depth gauge,* figure 16-13.

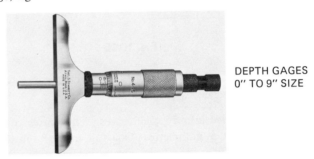

DEPTH GAGES
0" TO 9" SIZE

Fig. 16-13

How to Read a Micrometer — English System

The sleeve is divided into 40 even spaces. This corresponds to the pitch of the spindle which is 40 threads per inch. One complete turn of the thimble moves the spindle 1/40 or .025 inch, figure 16-14.

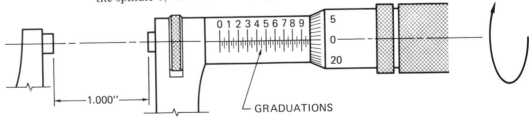

1.000"

GRADUATIONS

Fig. 16-14

The beveled edge of the thimble is divided into 25 equal parts. Each line equals .001 inch, figure 16-15.

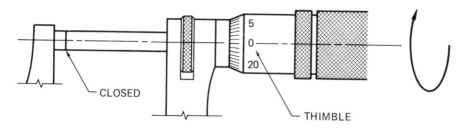

Fig. 16-15

To read the micrometer: Multiply the number of vertical lines on the sleeve by .025 inch. Add the number of thousandths indicated on the thimble, figure 16-16.

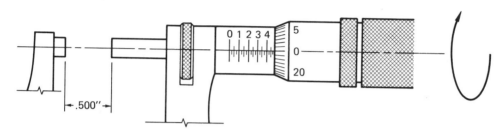

Fig. 16-16

How to Read a Micrometer — Metric System

The sleeve is divided into 50 even spaces. This corresponds to the pitch of the spindle which is 0.5 millimetre. One full turn of the thimble moves the spindle 0.5 millimetre, figure 16-17.

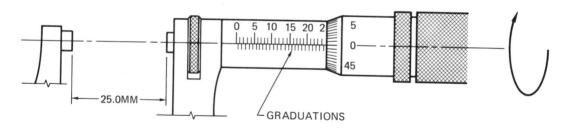

Fig. 16-17

The beveled edge of the thimble is divided into 50 equal parts. Each line equals 1/50 of 0.5 millimetre or 0.01 millimetre, figure 16-18.

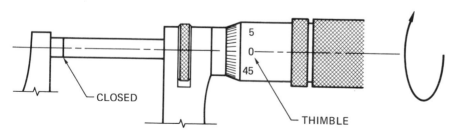

Fig. 16-18

To read the micrometer: Multiply the number of vertical lines on the sleeve by .05 millimetre. Add the number of millimetres indicated on the thimble, figure 16-19.

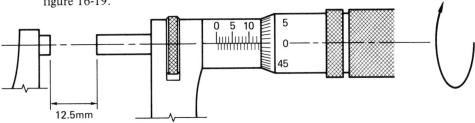

Fig. 16-19

SAMPLE READINGS

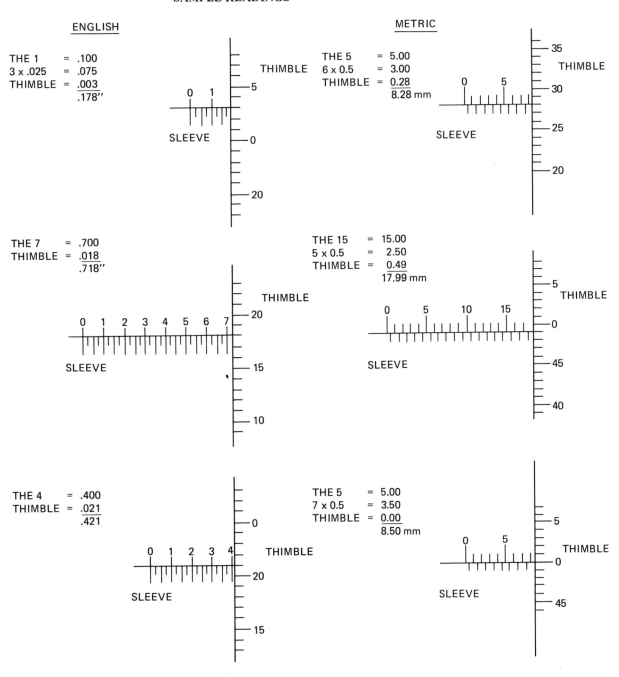

ENGLISH

THE 1   = .100
3 x .025   = .075
THIMBLE = .003
             .178″

THE 7   = .700
THIMBLE = .018
             .718″

THE 4   = .400
THIMBLE = .021
             .421

METRIC

THE 5   = 5.00
6 x 0.5   = 3.00
THIMBLE = 0.28
             8.28 mm

THE 15   = 15.00
5 x 0.5   = 2.50
THIMBLE = 0.49
             17.99 mm

THE 5   = 5.00
7 x 0.5   = 3.50
THIMBLE = 0.00
             8.50 mm

VERNIER CALIPERS

Vernier Calipers – English System

The vernier scale on the vernier caliper is read very much like a micrometer. The vernier, figure 16-20, reads to 1/1000 of an inch. The full inch is located on the main bar. Each inch is divided into four parts (.025 inch). The vernier scale is divided into 25 equal parts, each representing .001 inch.

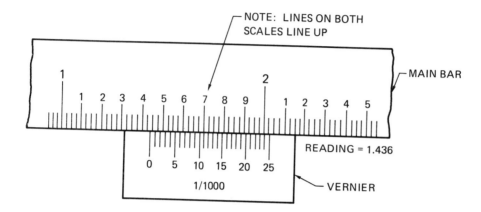

Fig. 16-20 Vernier caliper, English system

Using figure 16-20, a reading is calculated as follows:
From the main bar:

1. Record the full inches – 1.000 (in thousandths)
2. Record the 1/10th inch – .400 (in thousandths)
3. Record the 1/25th inch – .025 (in thousandths)

From the vernier scale:

4. Count the number
   of graduations on
   the vernier scale
   from 0 to a line
   that coincides with a
   line on the main bar   – .011 (in thousandths)
                              1.436 inches

Vernier Calipers – Metric System

The vernier scale in figure 16-21 is a metric venier and reads very much like the English vernier. It can be read to 0.02 millimetre. Each main bar graduation is in 0.5 millimetre with each 20th graduation numbered. The vernier is divided into 25 even parts, each representing 0.02 millimetre. To read the caliper, first record the total millimetres between zero on the main bar and zero on the vernier. Count the number of graduations on the vernier from zero to a line that coincides with a line on the main bar. Multiply that number times 0.02 millimetre and add this number to get the total reading.

Using the metric vernier scale in figure 16-21, the vernier zero is 41.5 millimetres past the location of the zero on the main bar. Therefore:

$$41.5 \text{ mm} + (9 \times 0.02 \text{ mm}) = 41.68 \text{ mm total reading.}$$

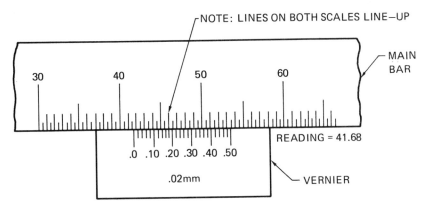

Fig. 16-21 Vernier caliper, metric system

Vernier calipers have the ability to measure both outside an object and inside an object, figures 16-22 and 16-23. When measuring an outside size, use the bottom scale. When measuring an inside size, use the top scale.

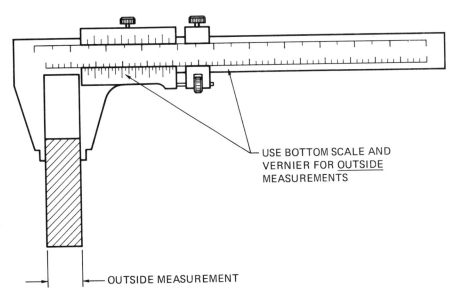

Fig. 16-22 Vernier caliper, outside measurement

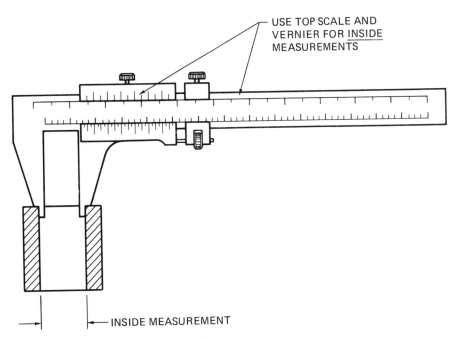

Fig. 16-23 Vernier caliper, inside measurement

UNIT 17

SPRINGS

SPRINGS

A manufacturing company often uses a standard spring in assembling products. Occasionally, however, a spring must be designed by a drafter for a special job. Figure 17-1 shows several types of springs.

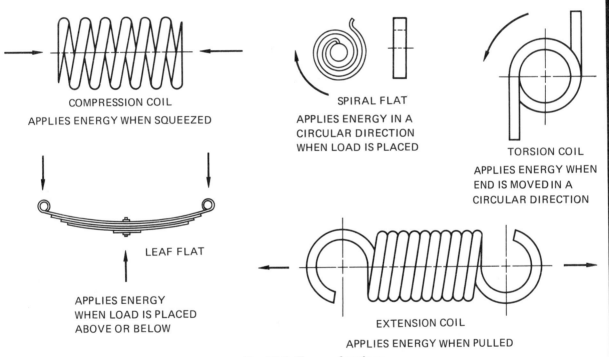

COMPRESSION COIL
APPLIES ENERGY WHEN SQUEEZED

SPIRAL FLAT
APPLIES ENERGY IN A
CIRCULAR DIRECTION
WHEN LOAD IS PLACED

TORSION COIL
APPLIES ENERGY WHEN
END IS MOVED IN A
CIRCULAR DIRECTION

LEAF FLAT
APPLIES ENERGY
WHEN LOAD IS PLACED
ABOVE OR BELOW

EXTENSION COIL
APPLIES ENERGY WHEN PULLED

Fig. 17-1 Types of springs

Study each part of the coil spring in figure 17-2.

Turn or *coil* is one complete turn about the center axis.

Total coils are the total number of turns or coils, starting from one point to the exact point on the next coil.

Active coils, in a compression spring, are usually the total coils minus the two end coils. Figure 17-2 has four active coils.

Free length is the overall length of the spring with no load on it. On a compression spring it is measured as noted in figure 17-2. On an extension spring, the measurement is taken from the inside of the hooks or ends.

Loaded length is the overall length of the spring with a given load applied to it.

Solid length, on a compression spring, is the length of the spring when it is completely compressed with each coil closed upon the next. In figure 17-2, if the wire diameter is .500 inch (12.7), the overall solid length is 1.5 inches (38).

Wire diameter is the diameter of the wire used to make the spring.

Outside diameter (O.D.) is the diameter of the outside of the coil spring.

Inside diameter (I.D.) is the diameter of the inside of the coil spring.

Mean diameter is the theoretical diameter of the spring measured to the center of the wire diameter. This diameter is used to draw the spring. To find the mean diameter, subtract the wire diameter from the outside diameter.

Direction of a spring describes whether the spring is wound left-hand or right-hand. Figure 17-2 is a left-hand spring.

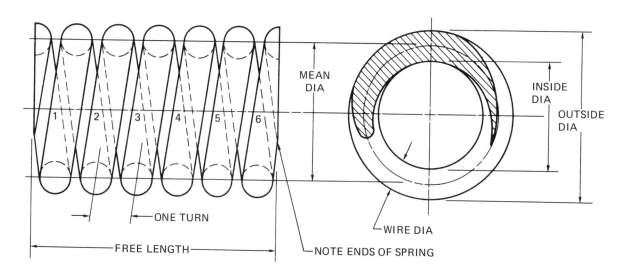

Fig. 17-2 Coil spring, detail drawing

Winding Direction

Springs can be wound left-hand or right-hand, figure 17-3. If the coil winding direction is not called out on a drawing, it will be manufactured with a right-hand winding.

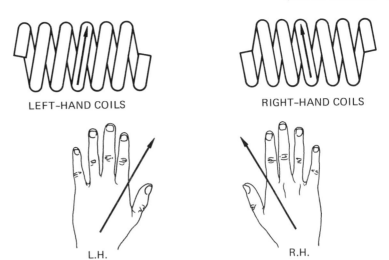

LEFT-HAND COILS   RIGHT-HAND COILS

L.H.   R.H.

Fig. 17-3 Determining direction of winding

COMPRESSION SPRING ENDS

Compression springs have either open ends, closed ends, ground open ends, or ground closed ends, figure 17-4. Note that the ground springs are made from plain open or closed springs. The plain open-ended spring is the most economical to manufacture, though its use is limited. Springs with ground closed ends are the most stable and can be ground back for even more support.

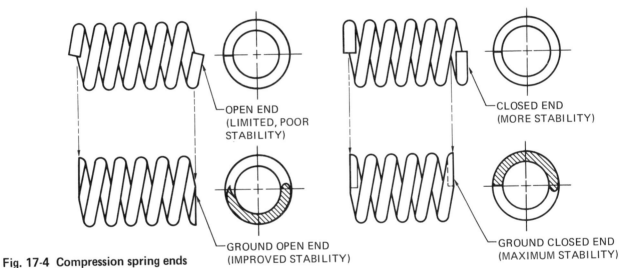

OPEN END (LIMITED, POOR STABILITY)

CLOSED END (MORE STABILITY)

GROUND OPEN END (IMPROVED STABILITY)

GROUND CLOSED END (MAXIMUM STABILITY)

Fig. 17-4 Compression spring ends

DRAWING A COMPRESSION SPRING

The following drawing procedure can be used to lay out a compression spring. As an example, a spring with these specifications is used:

Overall length 4″	L.H. winding
8 total coils (6 active)	1 1/2″ O.D.
Plain open ends	1″ I.D.
Wire size .25″	

Step 1. Lightly lay out the overall length and mean diameter of the spring. Divide the length into even spaces. The number of even spaces will depend on the total number of coils and the type of end required:

For plain open ends:	2 × total coils + 1
For plain closed ends:	2 × total coils − 1
For ground open ends:	2 × total coils − 1
For ground closed ends:	2 × total coils − 1

The spring in this example therefore has 17 even spaces (2 × 8 = 16 + 1 = 17).

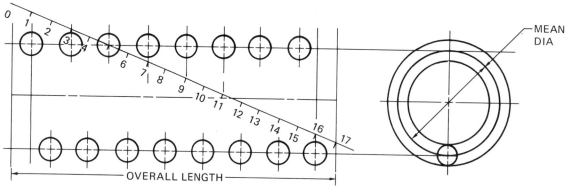

Step 2. Lightly draw circles at the top and bottom of each space to represent a cross section of each coil.

Step 3. Lightly draw the coil winding direction.

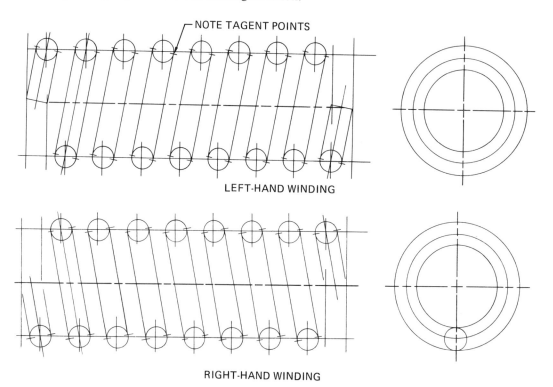

Step 4. Draw the style of end required. A plain open end is shown here. See figure 5-4 for other end styles.

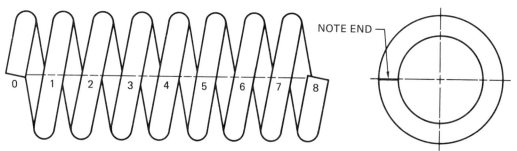

Step 5. Using correct line weight, finish the drawing. Add all dimensions and required notations.

EXTENSION SPRING ENDS

Figure 17-5 illustrates a few of the many types of ends used on *extension springs*. A loop completes a turn on itself while a hook is open. An extension spring has tight windings and is usually wound with right-hand windings.

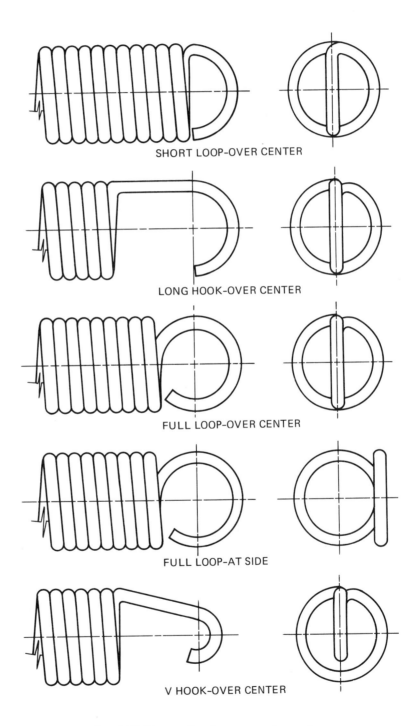

SHORT LOOP-OVER CENTER

LONG HOOK-OVER CENTER

FULL LOOP-OVER CENTER

FULL LOOP-AT SIDE

V HOOK-OVER CENTER

Fig. 17-5 Types of ends on extension springs

DRAWING AN EXTENSION SPRING

On a separate sheet of paper, draw a full loop over center extension spring following the steps given. Specifications:

    Overall length approx. 4.75″ (120 mm)
    Full loop, over center
    Wire size .188 (5)
    R.H. winding (standard)
    1.625″ (41) O.D.
    1.25″ (32) I.D.
    Extension spring style

Step 1. Make a rough sketch.

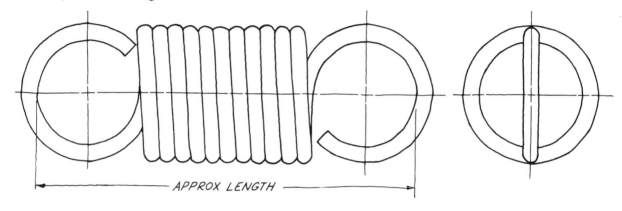

Step 2. Draw the end view (O.D./I.D.) and each end loop at the required length.

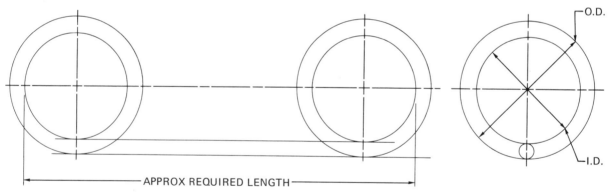

Step 3. Very carefully measure off the wire size springs and, with a circle template, lightly draw the approximate number of coils. Starting at the top from the right end, project to the left 1/2 wire size.

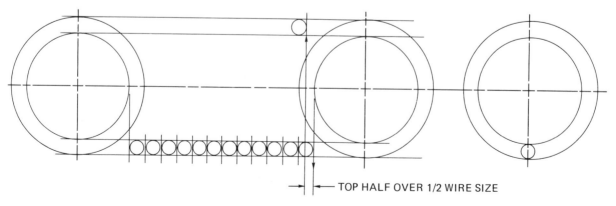

Step 4. Line up the drafting machine or triangle on the edge of the wire diameters (at X and Y) and lock on this angle. Lightly project up from wire diameters.

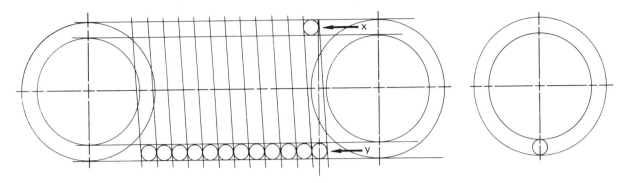

Step 5. Draw the wire diameters lightly. From the last wire size (A), project straight down. Adjust the end loop in. Note: Study the right end and be sure you understand how it is drawn (numbers 1, 2, 3, and 4).

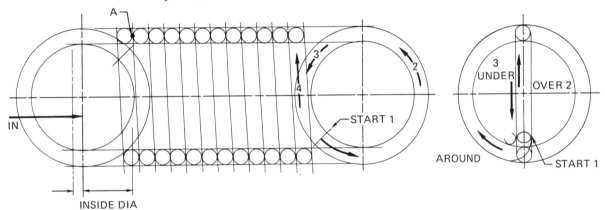

Step 6. Finish the drawing using correct line weight. Add dimensions and required notes.

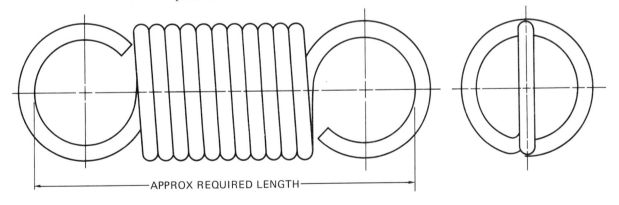

STANDARD DRAFTING PROCEDURES

Some companies will not pay a drafter to draw a conventional representation of a spring, figure 17-6, because of the time and cost involved.

A short cut method to draw a spring is shown in figure 17-7. This represents the spring in figure 17-6, but takes less time to draw. Even in semiconventional spring representations (schematics) all dimensions and notes must be added.

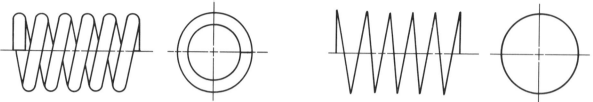

Fig. 17-6 Conventional spring drawing

Fig. 17-7 Schematic spring drawing

Another method of drawing long springs rapidly is shown in figure 17-8.

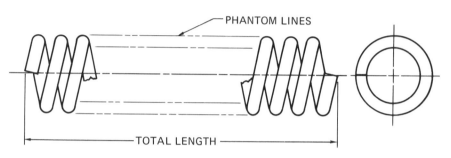

PHANTOM LINES

TOTAL LENGTH

Fig. 17-8 Drawing long springs rapidly

Figure 17-9 shows a section view of a small spring (top) and a large spring (bottom). Both are right-hand springs but, since the back half of the springs are shown, they appear left-hand in section.

SECTION VIEW OF A SMALL SPRING

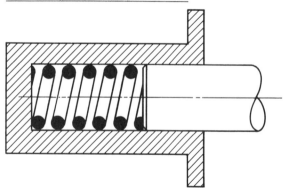

SECTION VIEW OF A LARGE SPRING

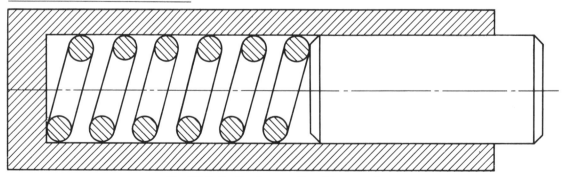

Fig. 17-9 Sectioning small and large springs

UNIT 18

CAMS

CAMS

A *cam* changes rotary motion. . . into up or down motion. . . .
The cam is attached to a rotating shaft. The up and down, or *reciprocating*,
motion is taken from the cam by the *follower*. Study the illustrations in figure
18-1. Note how the follower is at its highest point at 0 degree and 180 degrees and
at its lowest point at 90 degrees and 270 degrees.

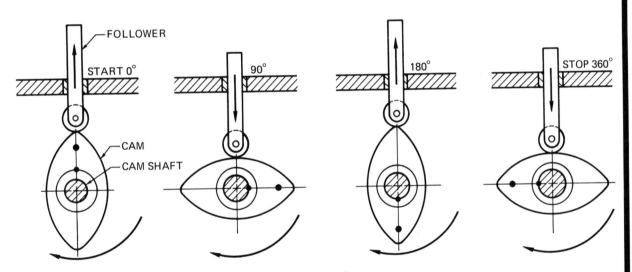

FOLLOWER

START 0°   90°   180°   STOP 360°

CAM
CAM SHAFT

Fig. 18-1 Reciprocating motion from rotary movement

BASIC TYPES OF CAM FOLLOWERS

In selecting a *cam follower,* speed of rotation and the various loads placed upon the lifter must be considered, figure 18-2.

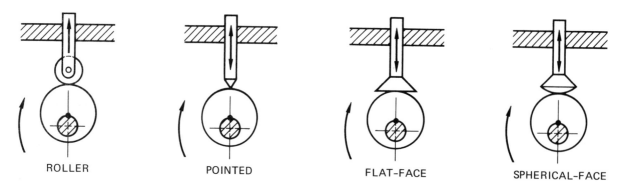

ROLLER   POINTED   FLAT-FACE   SPHERICAL-FACE

Fig. 18-2 Cam followers

Modified Cam Follower

The up and down motion can be modified by changing the rotary motion of shaft A, figure 18-3, into an up and down motion on follower B, and back into a rocker type motion on shaft C.

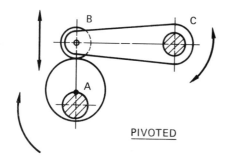

PIVOTED

Fig. 18-3 Modified cam follower

CAM MECHANISM

There are two major kinds of cams, *radial design* and *cylindrical design.* The radial design can be modified to get a rocker motion, figure 18-4.

FOLLOWER

CAM SHAFT

PERPENDICULAR

Ⓐ MOTION — ROTATION — CENTER OF SHAFT

Ⓑ MOTION — ROTATION — CENTER OF SHAFT

MODIFIED CAM

Fig. 18-4 Radial cam

Both designs use a *drive shaft* that rotates, but the action or direction of the followers differ. In radial arm design, the followers operate perpendicular to the cam shaft. In the cylindrical design, figure 18-5, the follower operates parallel to the cam shaft.

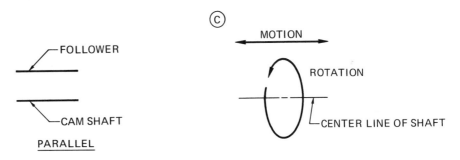

Fig. 18-5 Cylindrical cam

Figure 18-6 shows a cylindrical design. The shaft or rod that holds the follower does not rotate but moves back and forth. The follower produces movement parallel to the cam shaft.

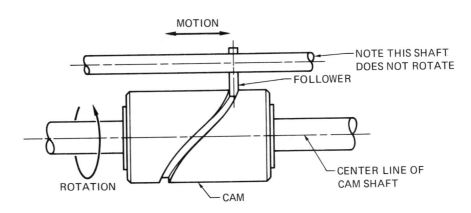

Fig. 18-6 Cylindrical cam. Follower produces movement parallel to cam shaft

DWELL

Dwell, or rest, is the time the cam follower is stopped or does not move up or down. The dwell is designed into the cam *profile*.

CAM DESIGN

In actual cam design, the cam follower type and location must be considered. For the drawing exercises in this unit, only cam design will be studied.

Study figure 18-7. Note that the cam rotates counterclockwise and the follower roller rotates *clockwise*. As the cam rotates, the follower will *drop* to location #2, then to #3, and so on. In designing a cam, it is important to consider the cam rotation direction. The cam is laid out and designed in the opposite direction of rotation.

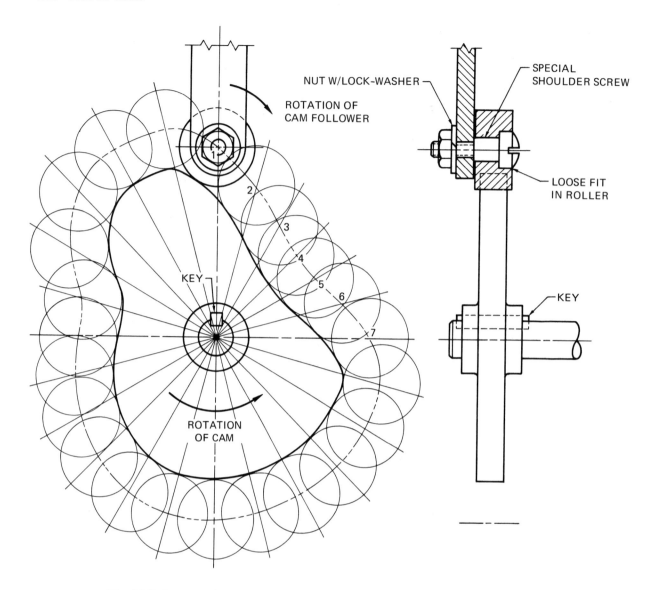

NUT W/LOCK-WASHER

ROTATION OF
CAM FOLLOWER

SPECIAL
SHOULDER SCREW

LOOSE FIT
IN ROLLER

KEY

KEY

ROTATION
OF CAM

Fig. 18-7 Cam rotation will cause the follower to drop and rise.

DISPLACEMENT DIAGRAMS

The *displacement diagram*, figure 18-8, is a curve that illustrates the exact motion of the follower through one full turn of the cam. The total overall length of the diagram does not have to be in scale or to exact dimension.

Terms of Displacement Diagrams

- *Length* equals one revolution of the cam. Length is drawn as a circumference of the work circle.
- *Working circle* is the radius equal to the distance from the center of the shaft to the highest point on the cam rise.
- *Height* is the maximum rise the follower travels. This is drawn to scale on the displacement diagram.
- *Time interval* is the time required for the cam to move the follower up or down.

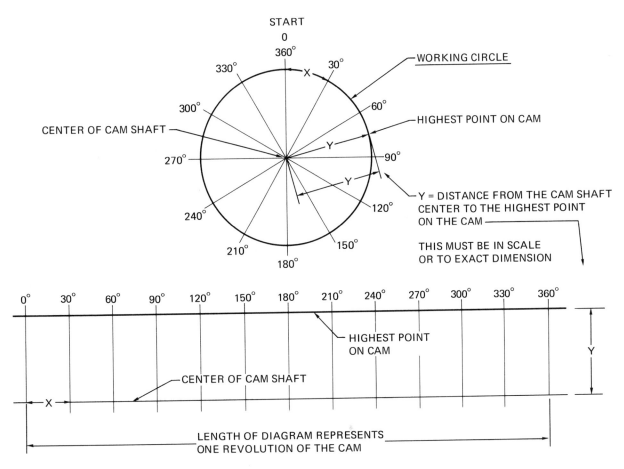

Fig. 18-8 Displacement diagram

CAM MOTIONS

There are three major kinds of cam motions; uniform velocity, harmonic motion, and uniform acceleration. Study how each is drawn.

Uniform Velocity

With uniform velocity, the follower rises and falls at a *constant speed*. This is very abrupt and rough, so it is usually modified by adding a 1/3 radius at the stop positions, figure 18-9.

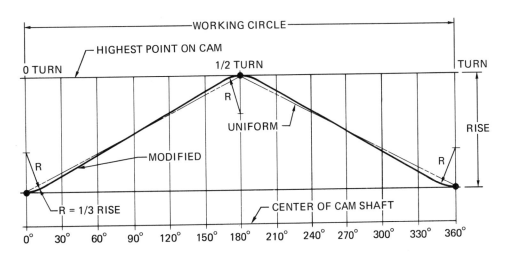

Fig. 18-9 Uniform velocity

Harmonic Motion

Harmonic motion is smooth, but the *speed is not uniform,* figure 18-10. It is used for high speed mechanisms as it has a smooth start and stop.

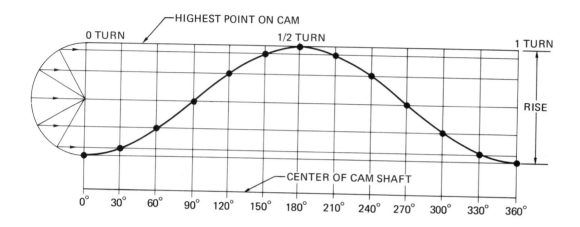

Fig. 18-10 Harmonic motion

Uniform Acceleration

Uniform acceleration is the smoothest of all cam motions. Its *speed is constant* throughout cam travel, figure 18-11.

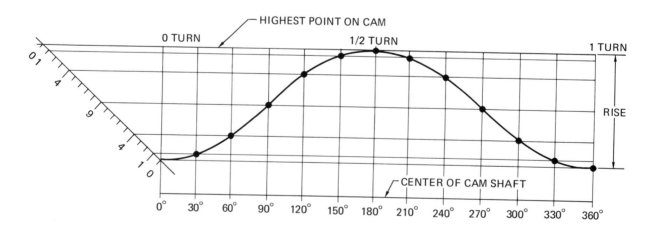

Fig. 18-11 Uniform acceleration

Notice how smooth the uniform acceleration cam is in comparison to the uniform velocity cam. Remember that the cam is always finished with an irregular curve in order to have a smooth cam profile. (Note: drawings are not full size)

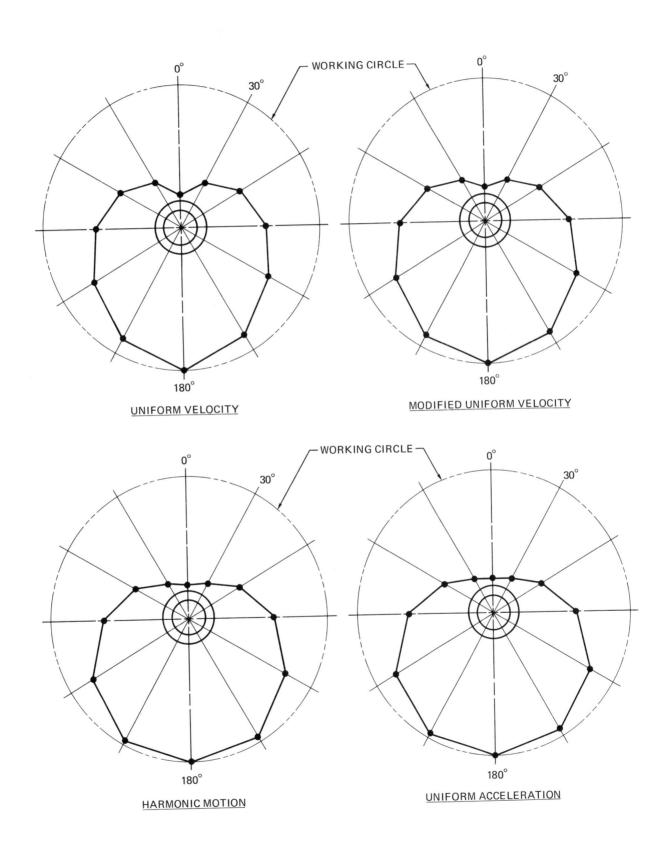

CAM DESIGN AND LAYOUT

Study how to draw a cam layout from the given requirements to the displacement diagram.

Given:

1. Rise 90°, modified uniform velocity, .75″ (19)
2. Dwell 15°
3. Rise 90°, harmonic motion, .75″ (19)
4. Dwell 15°
5. Fall 150°, uniform acceleration, 1.5″ (38)
6. Counterclockwise

Carefully lay out the displacement diagram, figure 6-12.

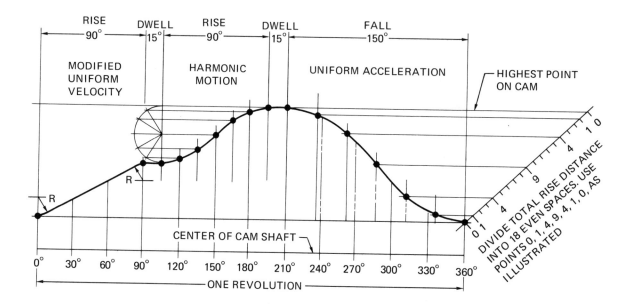

Fig. 18-12 Displacement diagram

Lay out cam. Always lay out cam points in directions opposite that of cam travel. In this example, figure 18-13, the cam rotates counterclockwise. Thus, lay out the cam clockwise as shown.

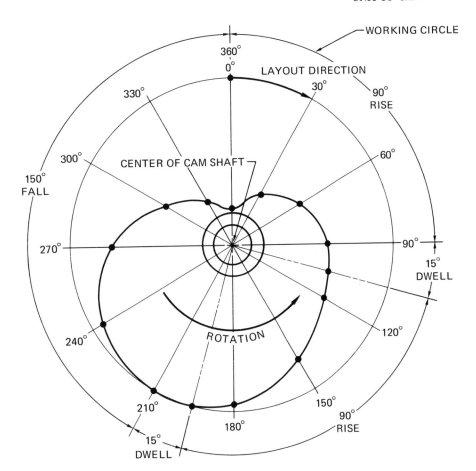

Fig. 18-13 Cam layout from given requirements

CAM TIMING

More than one cam is often attached to the same shaft. Each cam must function in relation to the other. Figure 18-14 shows the action of three cams attached to the same shaft.

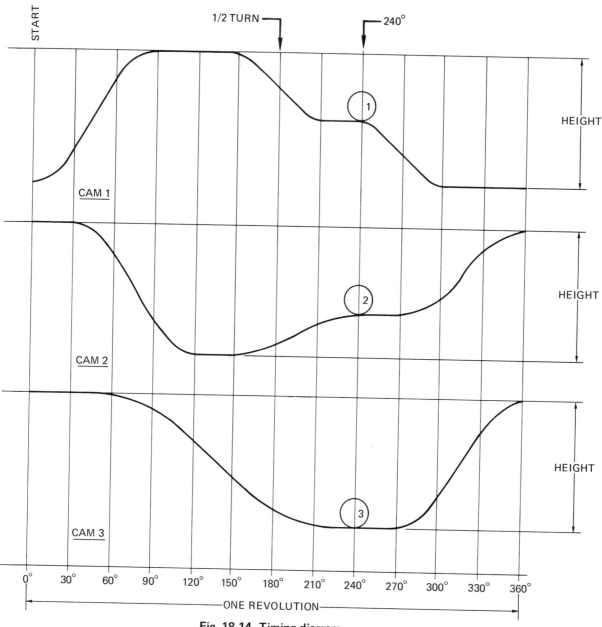

Fig. 18-14 Timing diagram

Notice cams 2 and 3 start at the highest point and cam 1 starts at the lowest point (0 degree). At 240 degrees, cam 1 is just completing dwell and is about to fall, cam 2 is just beginning a dwell, and cam 3 is halfway through a dwell. It is important these cams are designed with keyways so they cannot get out of time with each other.

DIMENSIONING CAMS

In order to dimension a cam, a displacement diagram must first be made to fit the requirements needed. The displacement diagram is then measured and dimensioned, figure 18-15.

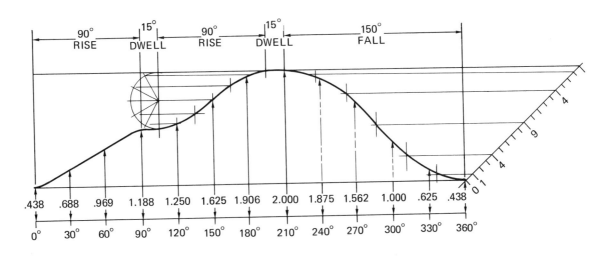

Fig. 18-15 Dimensioning displacement diagram

These dimensions are used to plot the points needed to lay out and dimension the cam. The cam in figure 18-16 is dimensioned from the center to the cam profile at every 30-degree segment of the working circle. Usually the rise, fall, and dwell areas are also included. This is only a suggested procedure as there is no standard for dimensioning cams. However, cams must be dimensioned neatly, completely, and accurately.

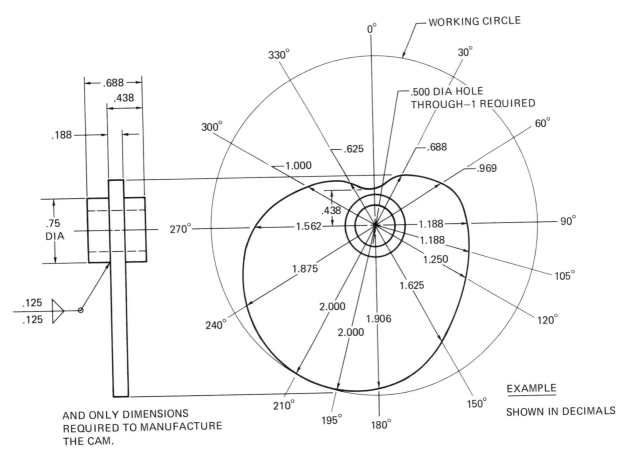

Fig. 18-16 Cam dimensioning

UNIT 19

ASSEMBLY & DETAIL DRAWING

KINDS OF DRAWINGS

There are various kinds of drawings used in a mechanical engineering department. The drafter must be able to recognize and draw each of them.

Design Layout Drawing

All major designs start from a design layout. This is usually a sketch drawing, full size or to scale. It is up to the drafter to draw each part depicted on this layout. In drawing the parts, the drafter must use the basic ideas in the design layout, but may change it to fit standard material stock, standard methods of manufacturing, and standard material sizes. If changes are made, they should be reviewed with the designer for his approval. Most of the time the designer draws parts as close to size as possible, but a design layout is not usually dimensioned unless particular dimensions must be maintained.

Assembly Drawing

Any product that has more than one part must have an assembly drawing. The assembly drawing shows how a product is assembled when completed. It can have one, two, three, or more views that are placed in the usual positions. One view is often a section view to illustrate the various parts and how they are assembled. Each part in an assembly drawing is identified by a circled detail number.

Permanently Fastened Parts Drawing

When two or more parts are permanently fastened such that they cannot be disassembled after assembly, they are called out as in a subassembly drawing.

Detail Drawing

Each part must have its own fully-dimensioned detail drawing, its own drawing number, and its own drawing title. All the information needed to manufacture the part is included in the detail drawing. The shape must be shown in the views given, features must be dimensioned and located, and specifications given on the drawing or title block.

Purchased Parts

A manufacturing company cannot afford to make standard items such as nuts, bolts, and washers which can be purchased ready for use. A drafter should try to design around standard parts whenever possible. Such parts are not drawn but simply called out on an assembly drawing by size, material, and finish. Other nonhardware parts may be treated in a similar fashion.

NOTE:

Study pages 269 through 277, identify each kind of drawing and how they relate to the assembly drawing on page 269.

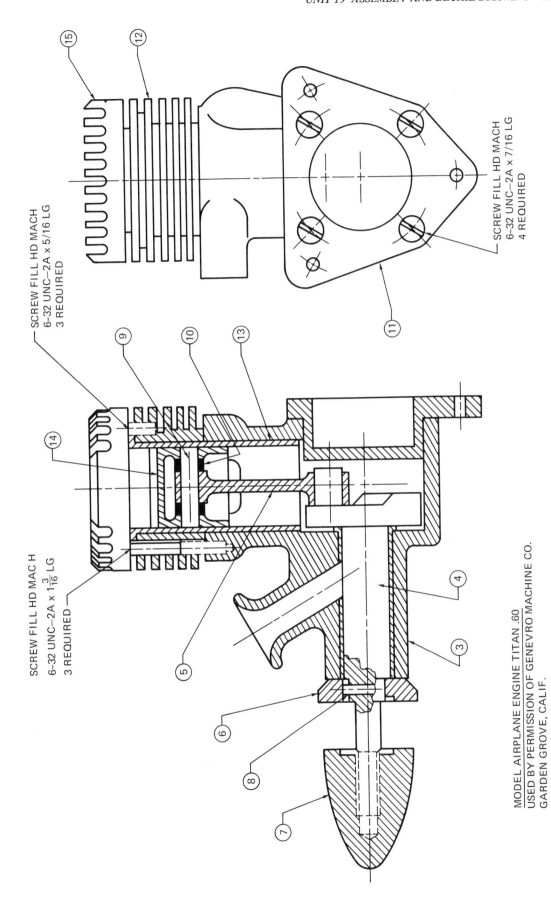

SCREW FILL HD MACH
6-32 UNC–2A x 5/16 LG
3 REQUIRED

SCREW FILL HD MACH
6-32 UNC–2A x 7/16 LG
4 REQUIRED

SCREW FILL HD MAC H
6-32 UNC–2A x 1 $\frac{3}{16}$ LG
3 REQUIRED

ASSEMBLY DRAWING

MODEL AIRPLANE ENGINE TITAN .60
USED BY PERMISSION OF GENEVRO MACHINE CO.
GARDEN GROVE, CALIF.

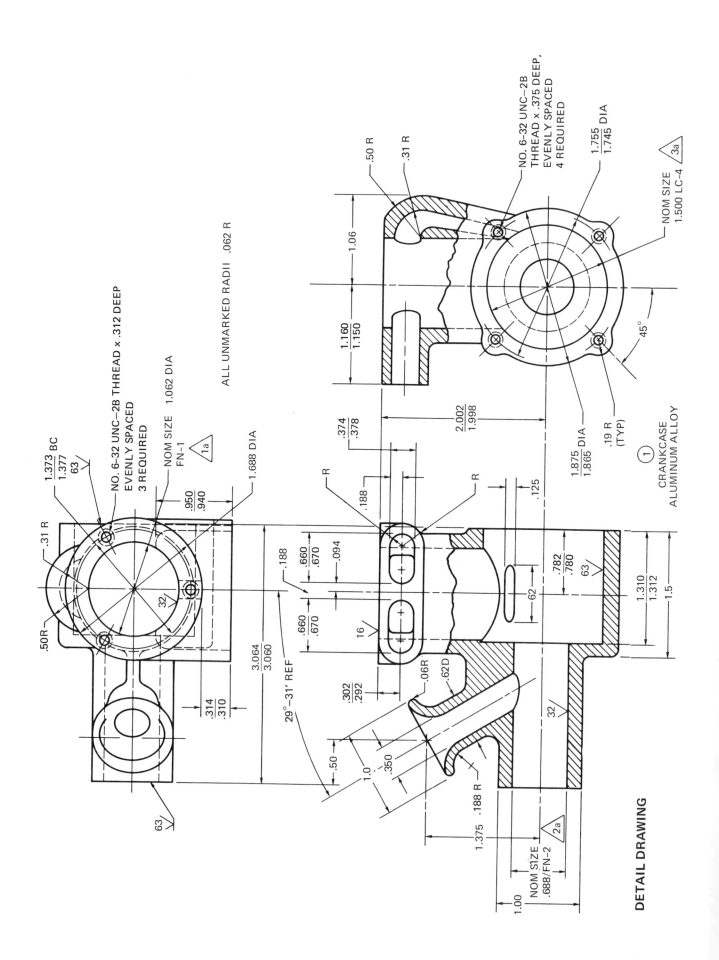

.50 R

.31 R

NO. 6-32 UNC–2B
THREAD x .375 DEEP,
4 REQUIRED

1.755 DIA
1.745

1.06

1.160
1.150

NOM SIZE
1.500 LC-4 ⟨3a⟩

45°

.374
.378

2.002
1.998

1.875 DIA
1.865

.19 R
(TYP)

.125

.188

R

R

.660
.670

.094

.782
.780

62

63

1.310
1.312

1.5

① CRANKCASE
ALUMINUM ALLOY

.660
.670

16

.302
.292

.06R

.62D

32

.50

1.0

.350

.188 R

2a⟩

1.375

NOM SIZE
.688/FN-2

1.00

DETAIL DRAWING

NO. 6-32 UNC–2B THREAD x .312 DEEP
EVENLY SPACED
3 REQUIRED

NOM SIZE 1.062 DIA
FN-1 ⟨1a⟩

ALL UNMARKED RADII .062 R

1.373 BC
1.377

63

.31 R

.50R

32

.314
.310

1.688 DIA

.950
.940

.188

3.064
3.060

29°–31' REF

63

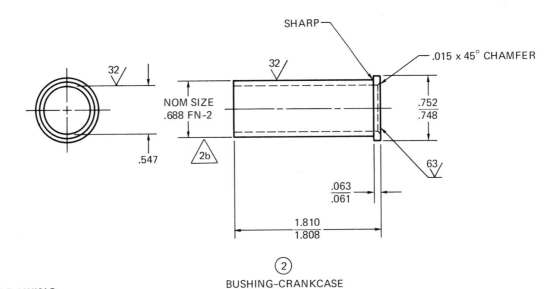

②
BUSHING–CRANKCASE
LEADED BRONZE

DETAIL DRAWING

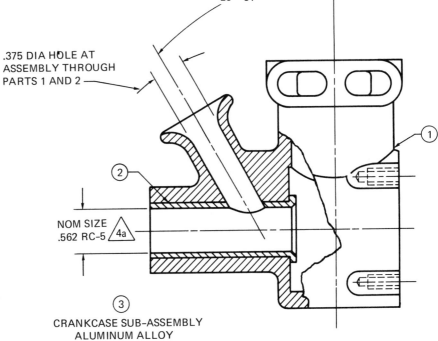

③
CRANKCASE SUB–ASSEMBLY
ALUMINUM ALLOY

SUB-ASSEMBLY

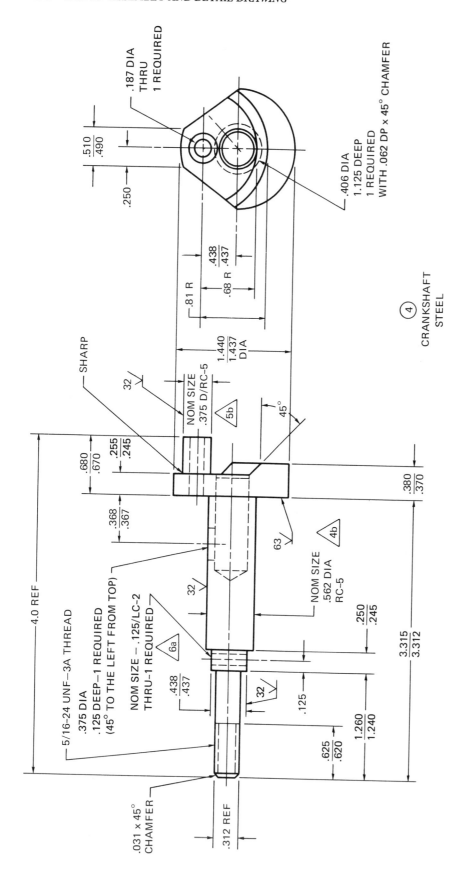

DETAIL DRAWING

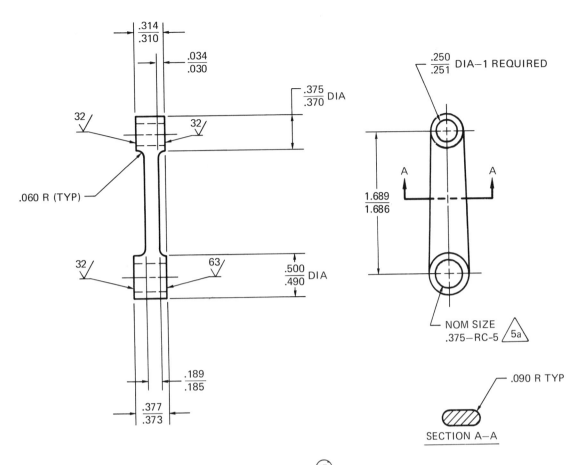

(5)
ROD-CONNECTING
ALUMINUM ALLOY

DETAIL DRAWING

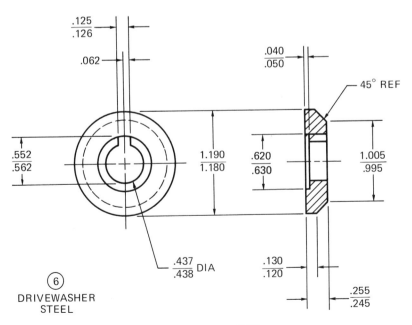

(6)
DRIVEWASHER
STEEL

DETAIL DRAWING

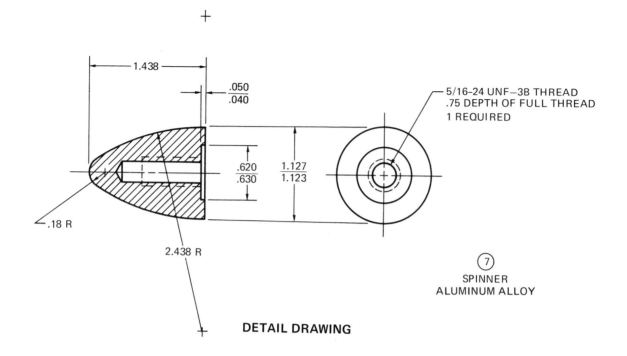

1.438

.050
.040

5/16-24 UNF—3B THREAD
.75 DEPTH OF FULL THREAD
1 REQUIRED

.620
.630

1.127
1.123

.18 R

2.438 R

⑦
SPINNER
ALUMINUM ALLOY

DETAIL DRAWING

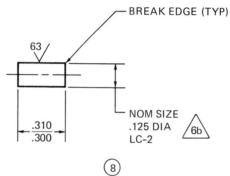

BREAK EDGE (TYP)

63

.310
.300

NOM SIZE
.125 DIA
LC-2

6b

⑧
PIN-DRIVEWASHER
(SCALE: DOUBLE SIZE)

DETAIL DRAWING

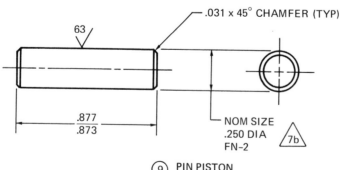

.031 x 45° CHAMFER (TYP)

63

.877
.873

NOM SIZE
.250 DIA
FN-2

7b

⑨ PIN PISTON
1010 STEEL
(SCALE: DOUBLE SIZE)

DETAIL DRAWING

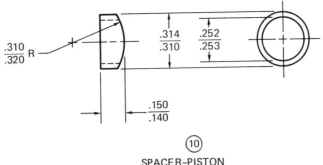

.310 / .320 R

.314 / .310

.252 / .253

.150 / .140

⑩
SPACER-PISTON
BRASS
(SCALE: DOUBLE SIZE)

DETAIL DRAWING

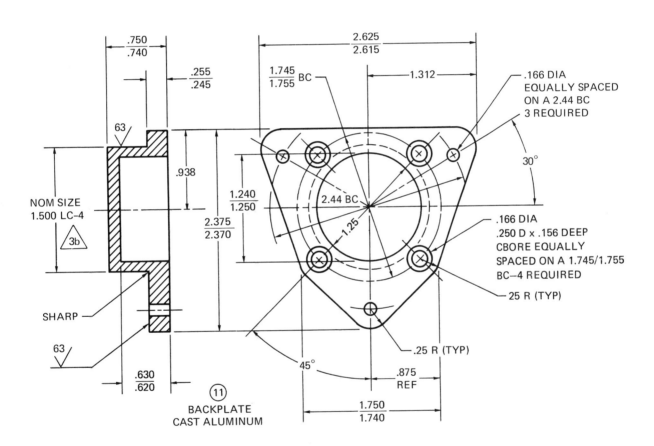

.750 / .740

.255 / .245

1.745 / 1.755 BC

2.625 / 2.615

1.312

.166 DIA
EQUALLY SPACED
ON A 2.44 BC
3 REQUIRED

.938

63

30°

NOM SIZE
1.500 LC-4

⚠ 3b

1.240 / 1.250

2.375 / 2.370

2.44 BC

1.25

.166 DIA
.250 D x .156 DEEP
CBORE EQUALLY
SPACED ON A 1.745/1.755
BC-4 REQUIRED

25 R (TYP)

SHARP

63

.630 / .620

⑪
BACKPLATE
CAST ALUMINUM

45°

.875
REF

.25 R (TYP)

1.750 / 1.740

DETAIL DRAWING

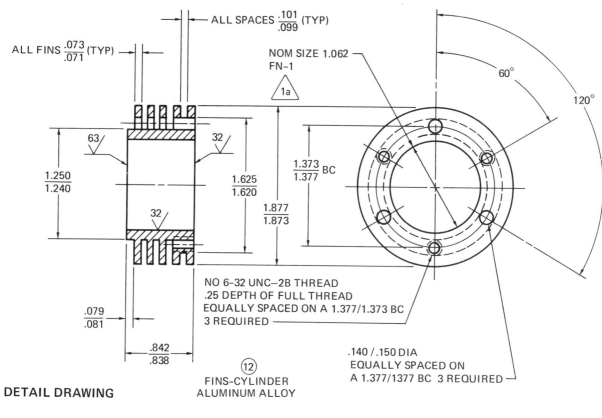

ALL SPACES $\frac{.101}{.099}$ (TYP)

ALL FINS $\frac{.073}{.071}$ (TYP)

NOM SIZE 1.062
FN-1

1a

60°

120°

63

32

1.250
1.240

1.625
1.620

1.373 BC
1.377

1.877
1.873

32

.079
.081

.842
.838

NO 6-32 UNC-2B THREAD
.25 DEPTH OF FULL THREAD
EQUALLY SPACED ON A 1.377/1.373 BC
3 REQUIRED

.140 /.150 DIA
EQUALLY SPACED ON
A 1.377/1377 BC 3 REQUIRED

DETAIL DRAWING

⑫
FINS-CYLINDER
ALUMINUM ALLOY

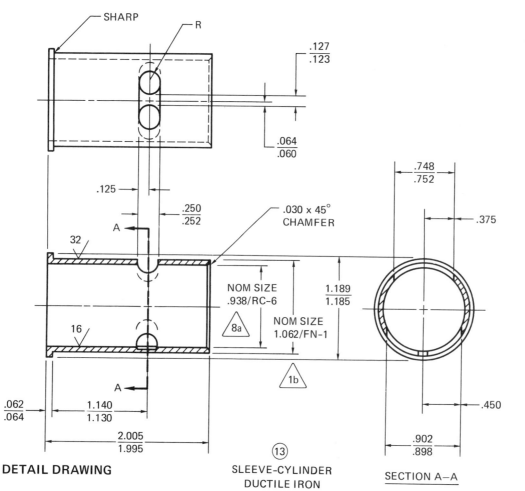

SHARP

R

.127
.123

.064
.060

.125

.250
.252

.030 x 45°
CHAMFER

.748
.752

.375

A

32

NOM SIZE
.938/RC-6

8a

1.189
1.185

NOM SIZE
1.062/FN-1

16

A

1b

.450

.062
.064

1.140
1.130

2.005
1.995

⑬

.902
.898

DETAIL DRAWING

SLEEVE-CYLINDER
DUCTILE IRON

SECTION A-A

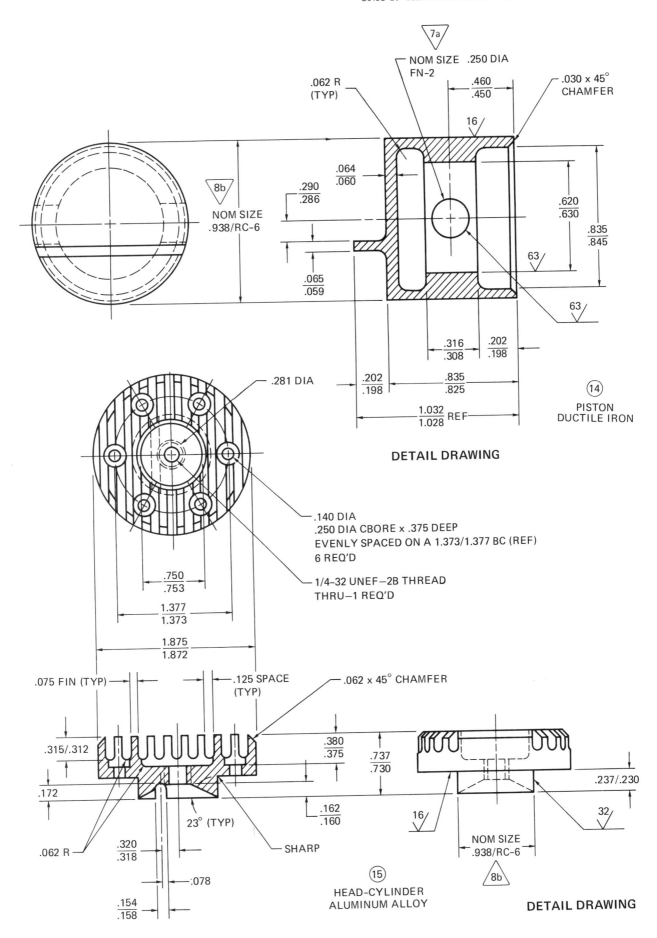

7a

NOM SIZE .250 DIA
FN-2

.062 R
(TYP)

.030 x 45°
CHAMFER

.460
.450

16 /

.064
.060

8b

NOM SIZE
.938/RC-6

.290
.286

.620
.630

.835
.845

63 /

.065
.059

63 /

.316
.308

.202
.198

.202
.198

.835
.825

14

PISTON
DUCTILE IRON

1.032
1.028 REF

DETAIL DRAWING

.281 DIA

.140 DIA
.250 DIA CBORE x .375 DEEP
EVENLY SPACED ON A 1.373/1.377 BC (REF)
6 REQ'D

1/4-32 UNEF-2B THREAD
THRU-1 REQ'D

.750
.753

1.377
1.373

1.875
1.872

.075 FIN (TYP)

.125 SPACE
(TYP)

.062 x 45° CHAMFER

.315/.312

.380
.375

.737
.730

.237/.230

.172

.162
.160

16 /

32 /

23° (TYP)

SHARP

NOM SIZE
.938/RC-6

8b

.062 R

.320
.318

15

HEAD-CYLINDER
ALUMINUM ALLOY

:078

.154
.158

DETAIL DRAWING

UNIT 20

MECHANICAL LETTERING

MECHANICAL LETTERING

Mechanical lettering can be done with a template and scriber, a lettering typewriter, or by using pressure-sensitive letters. Figure 20-1 illustrates one method of mechanical lettering.

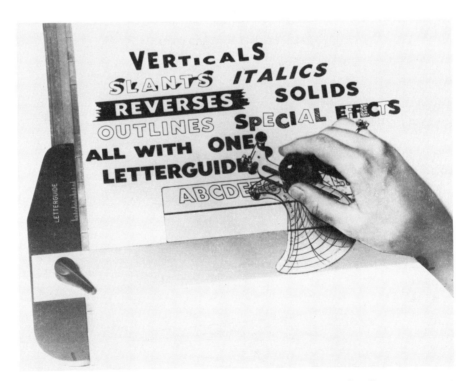

Fig. 20-1 Mechanical lettering with template and scriber

SCRIBER TEMPLATES

Scriber templates consist of laminated strips with engraved grooves to form letters. A *tracer pin* moving in the grooves guides the scriber pen or pencil in forming the letters, figure 20-2.

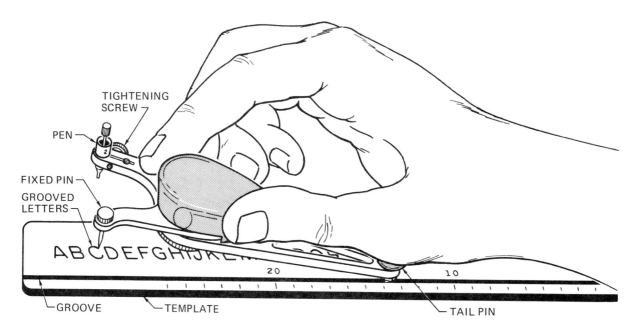

USE A VERY LIGHT DELICATE TOUCH.

Fig. 20-2 Scriber and template parts

Guides for different sizes and kinds of letters are available for any of the lettering devices. Different point sizes are made for special pens so that fine lines can be used for small letters and wider lines for larger letters. Scribers may be adjusted to get vertical or slanted letters of several sizes from a single guide by simply unlocking the screw under the scriber and extending the *arms*, figures 20-3 and 20-4.

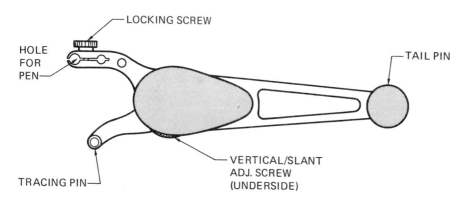

Fig. 20-3 Scriber for vertical letters

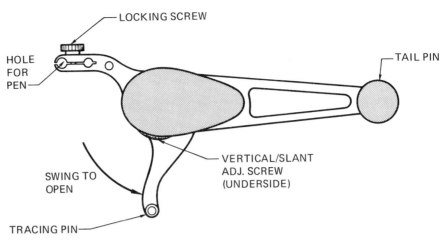

Fig. 20-4 Scriber for slant letters

One of the principal advantages of lettering guides is in maintaining uniform lettering, especially where there are a number of drafters. Another important use is for lettering titles, note headings, and numbers on drawings and reports.

Letters used to identify templates are:

U = *UPPERCASE*
L = *LOWERCASE*
N = *NUMBERS*

Thus a template 8-ULN means it is 8/16 inches high (1/2") and has uppercase and lowercase letters and numbers.

Tracing Pin

Better, more expensive scribers use a double tracing pin. The blunt end is used for single-stroke lettering templates or very large templates which have wide grooves. The sharp end is used for very small lettering templates, double-stroke letters, or script-type lettering using a fine groove. Most tracing pins have a sharp point, but some do not, figure 20-5. Always screw the cap back on after turning the tracing pin. Be careful with the points as they will break if dropped and can cause a painful injury if mishandled.

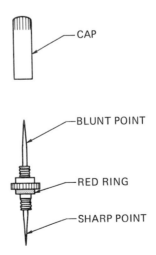

Fig. 20-5 Tracing pin (fixed pin)

INK

Always use fresh ink in the pens. In warm weather the ink will plug the pen if it is not changed weekly. Always clean the ink out of the pens if they will not be used for three or four days.

LETTERING SET

One *Leroy® lettering set* is designed for the beginning illustrator. The set contains the basic components needed to do controlled lettering, figure 20-6.

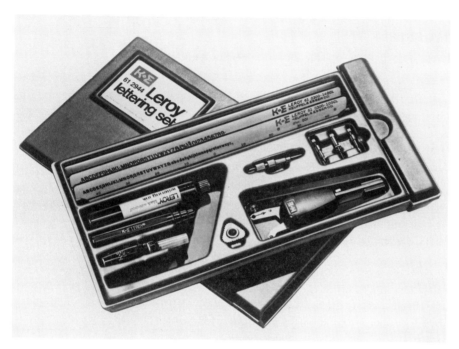

Fig. 20-6 A Leroy® lettering set

STANDARD TEMPLATE

Learning to form mechanical letters requires a great deal of practice. Figure 20-7 shows a template having three sets of uppercase and lowercase letters. Practice forming each size letter and number until they can be made rapidly and neatly. Use a very light, delicate touch so as not to damage the template, scriber, or pen.

Fig. 20-7 Template

SIZE OF LETTERS

The size or height of the letter is called out by the number used to identify each set. Sizes are in thousandths of an inch. A #100 is 0.100 inch high or slightly less than an eighth of an inch, while a #240 is 0.240 inch high, or slightly less than a quarter of an inch.

There is another system of template sizes which use simple numbers. These numbers are placed over 16 to give the height of the letter. The number 3, for instance, would be 3/16 inch in height.

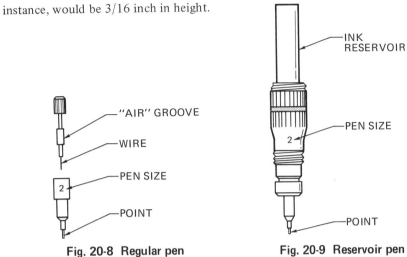

Fig. 20-8 Regular pen   Fig. 20-9 Reservoir pen

PENS

There are two types of pens, the regular pen and the reservoir pen. The regular pen must be cleaned after each use. The reservoir pen should be cleaned when it gets "sluggish" or before being stored for long periods of time, figures 20-8, 20-9, and 20-10.

Fig. 20-10 Technical pens. Note points. Top: pen is used for lines only. Bottom: pen is used for lines and in lettering scribers.

Figure 20-11 shows the recommended pen sizes for given letter heights.

Height of letter	Recommended Pen Size
.080	0 0 0
.100	0 0
.120	0
.140	1
.175	2
.200	3
.240	3
.290	4
.350	4
.425	5
.500	6

Fig. 20-11

Point Sizes

There are fourteen (14) point sizes ranging from #000, which is very thin and very delicate, through #14, which is very thick, figure 20-12. Extreme care must be used in replacing the fine wire into the body of the pen.

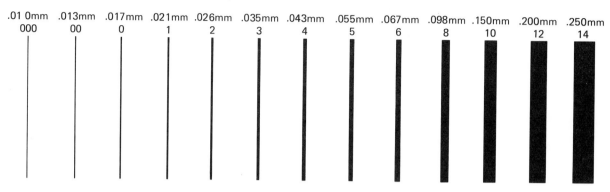

| .01 0mm | .013mm | .017mm | .021mm | .026mm | .035mm | .043mm | .055mm | .067mm | .098mm | .150mm | .200mm | .250mm |
| 000 | 00 | 0 | 1 | 2 | 3 | 4 | 5 | 6 | 8 | 10 | 12 | 14 |

Fig. 20-12 Lines drawn by various point sizes

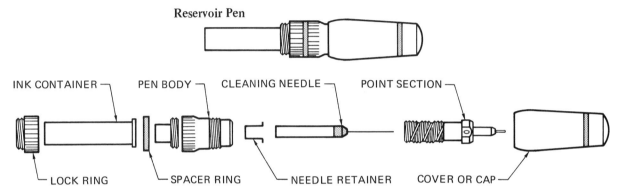

Reservoir Pen

INK CONTAINER — PEN BODY — CLEANING NEEDLE — POINT SECTION —

— LOCK RING — SPACER RING — NEEDLE RETAINER — COVER OR CAP —

Fig. 20-13 Reservoir pen parts

Reservoir Pen

Cleaning. Pens can be ruined by improper cleaning. Study steps one through five and follow them closely when cleaning pens, figure 20-13.

1. Remove cap and ink container.
2. Soak the body of the pen in hot water. The ink container should also be soaked if ink has dried in it.
3. After soaking, remove the pen body from the water. Hold the knurled part of the body with the top downward. Unscrew and remove point section. Remove the end of the cleaning wire weight. Do not bend the cleaning wire or it will break.
4. Immerse all body parts in a good pen cleaning fluid or hot water mixed half with ammonia.
5. Dry and clean.

Filling. To fill the pen, follow steps one through five:

1. Unscrew and remove knurled lock ring.
2. Remove ink container. Leave spacer ring in place.
3. Fill ink container with lettering ink. Do not fill more than 3/4 inch from top.
4. Hold filled container upright and insert pen body into container.
5. Replace knurled lock ring.

Use. Remove the cap from the pen body. To start the flow of ink, force the point downward and shake rapidly against a cloth. The pen is ready for use when ink appears on the cloth. Filled pens should be capped and kept in a vertical position with their tips upward when not in use.

JUSTIFYING COLUMNS

Printed text in newpapers and magazines are arranged in columns of equal width. Each line of type starts at the left-hand margin and ends at the right-hand margin. All of the lines are equal in length. The process of obtaining printed columns with lines of equal length is called *justification.*

BUTTERFLY-TYPE SCRIBER

Basic Parts

The *butterfly-type scriber* shown in figure 20-14 is a delicate, precision tool that will do its job without requiring any adjustments, repairs, or maintenance.

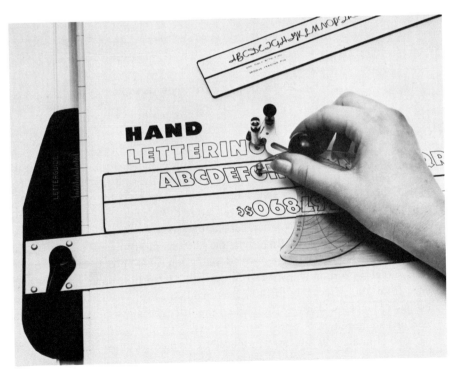

Fig. 20-14 A butterfly-type scriber

The vinyl base of the scriber bears the setting chart used in adjusting the pen arm for enlargements, reductions, verticals, and slants to be produced by tracing the engraved letters of a letter guide template.

The pen arm of the scriber holds the pen accessories for the various jobs to be performed. The pen and the arm have a thumb tightening screw device for securing the pen being used, and an adjustable pressure post screw with locking nut for controlling the amount of pressure and/or depth the pen or knife is set at, figure 20-15. The pressure post rides on the surface of the work when in use and is used only in conjunction with the swivel knife. The "bull's-eye" setting marker at the opposite end of the pen arm offers a concise, accurate means of setting the scriber for the various percentages and angles desired.

The tracing pin is the hardened tool steel point used in tracing the template letter. The tail pin serves as the pivot point for the triangular action of the scriber. This pin travels in the center groove of the template.

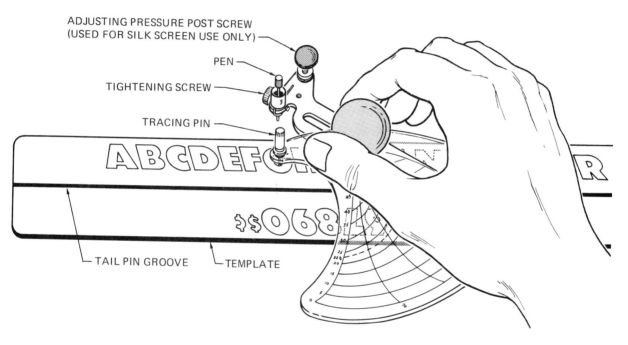

ADJUSTING PRESSURE POST SCREW
(USED FOR SILK SCREEN USE ONLY)

PEN

TIGHTENING SCREW

TRACING PIN

TAIL PIN GROOVE   TEMPLATE

Fig. 20-15 Using a butterfly-type scriber

Operation

This precision lettering tool is the key to producing clean, sharp, controlled lettering. The setting chart, using the bull's-eye at the end of the pen arm for a marker, begins at the outer edge with a starting line marked "vertical." In this position the scriber produces a vertical letter of normal size from the template being used. To enlarge this letter, set the bull's-eye at a position above the 100 percent intersection. At 120 percent, the scriber will produce a letter 20 percent greater in height than at 100 percent. A reduction can be produced by setting the bull's-eye at a position below the 100 percent intersection. Variations in height range from 100 percent up to 140 percent and down to 60 percent. The extreme settings produce condensed letters while the intermediate settings produce either headings, subheadings, or large or small letters.

Slants in all sizes are easily produced by setting the bull's-eye on a line other than the vertical line. Normal slants or italics are produced in all height adjustments by setting the bull's-eye on either the 15-degree or 22 1/2-degree line, and at the desired percentage of height of the letter on the template. Variations may be produced in slants ranging from 0 degrees to 50 degrees forward.

Tracing the engraved template letter requires a very light and delicate touch. This results in more accurately traced letters and less wear on the equipment. Each lettering application will require its own specific pen and will place at the fingertips of the drafter the very best in standard typeface and hand-lettered alphabets for fast, easy rendering.

LARGE LETTER SERIES LETTER GUIDES

Scriber settings differ for all letter guide templates bearing letters larger than number 12. For these large letter series letter guides, start with the bull's-eye set on the dotted line marked 22 1/2 degrees for all vertical letters. The intersection of this 22 1/2-degree line with the 100 percent line marks the setting for the normal size vertical letter from the template being used. All size 16, 20, 26, and 32 letter guides produce slants and reductions. Enlargements and some back-slants are possible with all sizes except 32. Two-inch lettering is the maximum size possible.

Fig. 20-16 Sample letters using a Letter Guide Collegiate-A Template

LETTER SPACING

Fast and optically correct spacing requires practice. After starting the heading by tracing the first letter, place the tracing pin of the scriber in the left-hand side of the next letter, slide the template and scriber into the proper optically spaced position, and proceed to trace. Perfect optical spacing can be obtained, even when lettering such letter combinations as AV, LA, YA, etc., by first planning the heading as normal, close, or wide space. Headings can be letter spaced to fit any required word length.

BALL-POINT PEN

A ball-point pen is an excellent tool for producing fast, sharp, concise lettering. The sealed-in ink supply is adequate for tracing upwards of 1000 letters.

CENTERING

ST. JOHNSBURY ACADEMY —120%

MAIN ST. —100%

ST. JOHNSBURY, VERMONT —60%

05819 —80%

Fig. 20-17 Sample letters with heights (step 1) and centers (step 2) indicated

Step 3. Cut the lettering into strips of paper, figure 20-18.

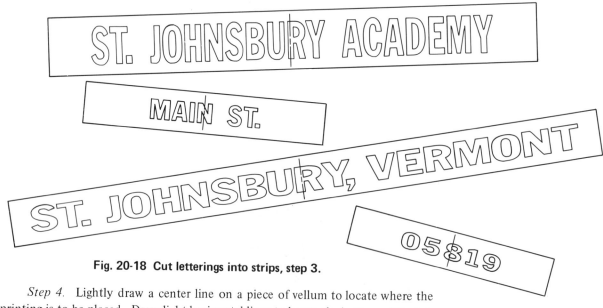

Fig. 20-18 Cut letterings into strips, step 3.

Step 4. Lightly draw a center line on a piece of vellum to locate where the printing is to be placed. Draw light horizontal lines to locate the bottom of the lettering, figure 21-19.

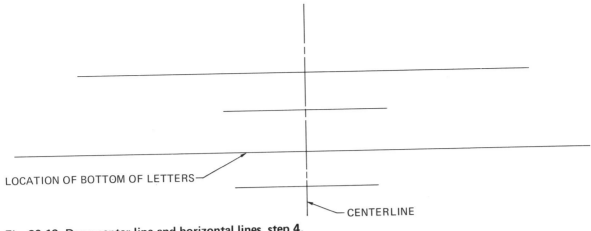

LOCATION OF BOTTOM OF LETTERS

CENTERLINE

Fig. 20-19 Draw center line and horizontal lines, step 4.

Step 5. Tape the cut strips under the vellum, lining up the center line on the paper with the center line on the strips. Line up the bottom of the words with the horizontal lines drawn in step 4. Fasten as illustrated, figure 20-20, using a minimum amount of tape.

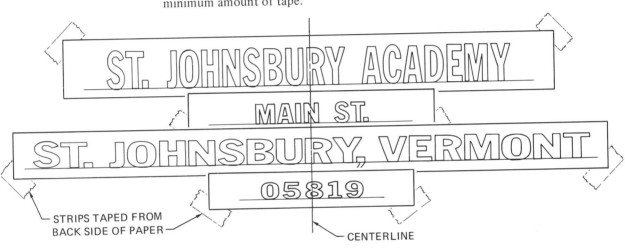

STRIPS TAPED FROM
BACK SIDE OF PAPER

CENTERLINE

Fig. 20-20 Fasten strip under vellum, step 5.

Step 6. Study the layout. If it appears centered and has no errors, reprint (trace) the words. Fill in lettering neatly. Recheck again. Note that when lettering on a card or using a border, always leave a little more space at the bottom of the lettering than at the top, figure 20-21.

ST. JOHNSBURY ACADEMY
MAIN ST.
ST. JOHNSBURY, VERMONT
05819

Fig. 20-21 Letter centering completed

UNIT 21

PERSPECTIVE DRAWING

ISOMETRIC PERSPECTIVE

Figures 21-1, 21-2, and 21-3 are examples of a *three-view drawing*, an *isometric drawing*, and a *perspective drawing* of the same object. The three-view drawing is used by the craftsman to make the object. The isometric and perspective drawings are types of pictorial representations to make interpreting the three-view drawing easier.

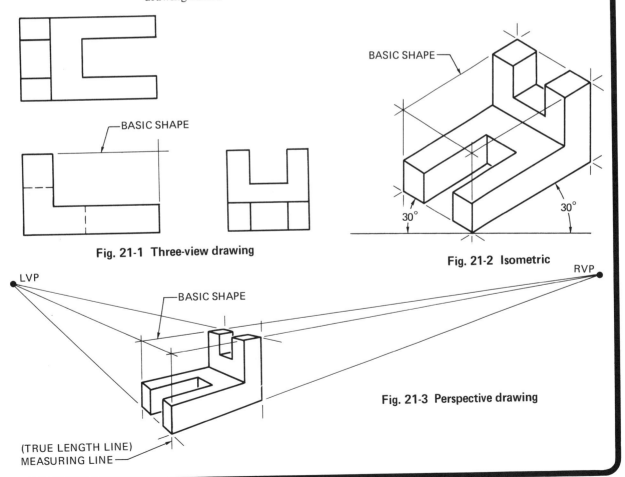

BASIC SHAPE

Fig. 21-1 Three-view drawing

BASIC SHAPE

30°   30°

Fig. 21-2 Isometric

LVP

RVP

BASIC SHAPE

Fig. 21-3 Perspective drawing

(TRUE LENGTH LINE)
MEASURING LINE

As previously explained, isometric drawings are made around an *isometric axis*, figure 21-4. All true measurements are laid out on these axis lines consisting of 30-degree angular lines projected to the left and right of center, and a vertical line projected upwards from center. The finished isometric gives a satisfactory pictorial drawing but is somewhat distorted.

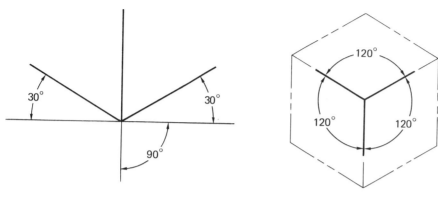

Fig. 21-4 An isometric axis

Perspective drawings are drawn with all true measurements on the true length line, see figure 21-3. Heights are drawn from that line to the vanishing points. This type drawing produces *pictorial drawings* which appear like actual photographs.

ONE-POINT PERSPECTIVE

One-point perspective is the simplest perspective to draw. Take a standard front view of an object and extend all other lines to one vanishing point (VP). Other variations are possible, figure 21-5. Boxes A and B show the bottom as well as one side because they are *above* eye level. Boxes C and D show one side but neither the top nor bottom because they are located *on* eye level. Boxes E and F show one side and the top because they are located *below* eye level.

Care should be used in selecting one point perspective views as they can be so close to pictorial views (A, B, E, F) that a pictorial view might actually be a

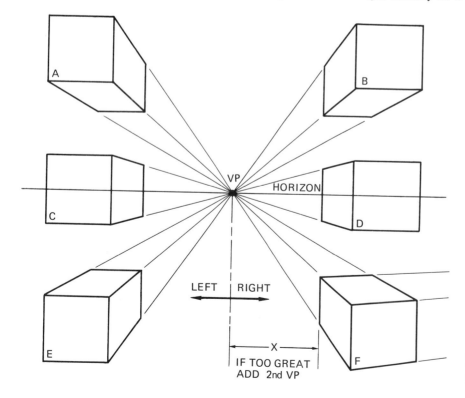

Fig. 21-5 Various layouts using one-point perspective

better choice and a great deal easier to draw. The main difference will be that receding lines will be parallel in pictorial views rather than converge. Also, if the object is placed too far to one side of the vanishing point (X), distortion occurs and a second vanishing point should be used.

Fig. 21-6 Objects drawn using one-point perspective

TWO-POINT PERSPECTIVE

Two-point perspective is similar to one-point perspective except that edges are projected to two points instead of one. Study the drawings which result by placing the object above, on, and below the horizon line, or eye level, in figure 21-7. The bottom of the object is visible when it is placed above the horizon line. Neither the top nor bottom of the object can be seen when the object's center is placed on the horizon line. The top is visible when the object is blaced beneath eye level.

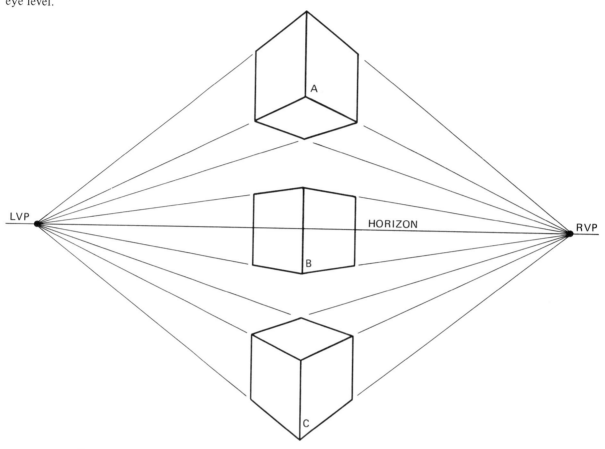

Fig. 21-7 An object drawn in two-point perspective on, above, and beneath eye level

Sketch as many objects as possible using this method. Two-point perspective is used more frequently than other methods and will be used to complete assignments in this unit.

Illustrated in figures 21-8 and 21-9 are two uses of two-point drawings. Figure 21-8 is an outside view of a building using two-point perspective. Notice how the

Fig. 21-8 Outside view in two-point perspective

windows get closer together and smaller as they go back in space, giving the impression of distance. The same is true of the inside view of a room using two-point perspective in figure 21-9.

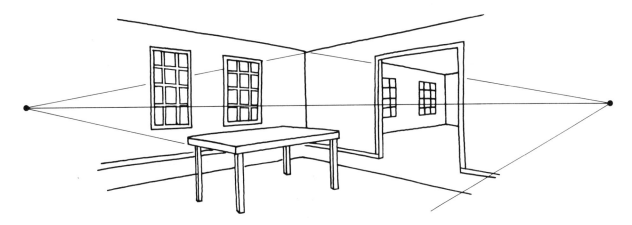

Fig. 21-9 Inside view in two-point perspective

THREE-POINT PERSPECTIVE

Three-point perspective is more difficult to draw than the other methods covered and, for this reason, is not often used. There are no parallel edges in a three-point drawing as all edges converge to one of the three vanishing points. Figure 21-10 shows an object drawn beneath the horizon. As in one- and two-point perspectives, objects using the three-point method can be placed on, above, or below the horizon line depending on the view desired.

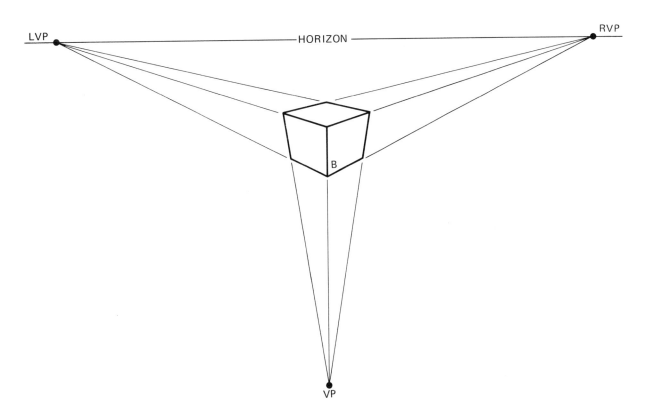

Fig. 21-10 Object drawn using three-point perspective

Three-point perspective is used only in special cases. Practice three-point perspective sketches by viewing the same objects from different positions, figure 21-11.

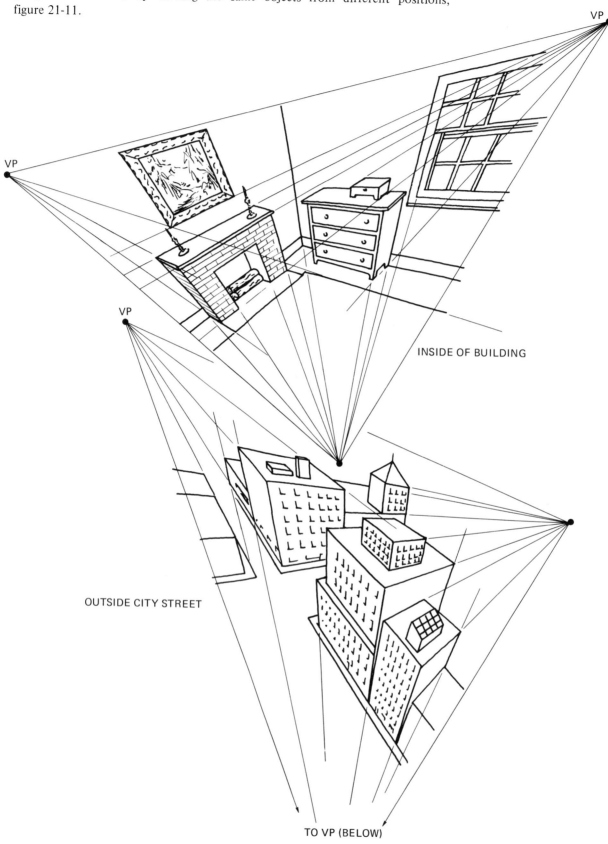

VP

VP

VP

INSIDE OF BUILDING

OUTSIDE CITY STREET

VP

TO VP (BELOW)

Fig. 21-11 Example of three-point perspective

PERSPECTIVE DRAWING TERMS

Figure 21-12 illustrates the terms used in perspective drawing. Refer to the illustration as the terms are defined.

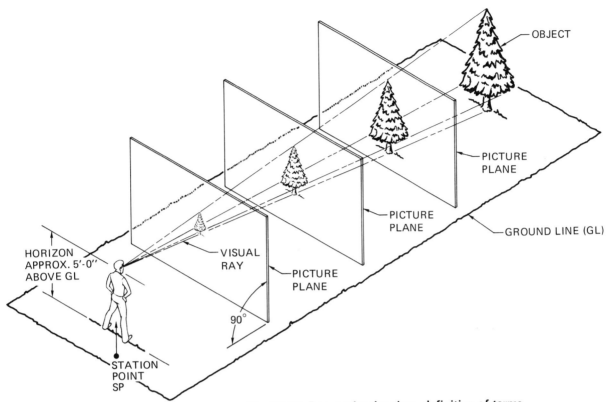

Fig. 21-12 Perspective drawing: definition of terms

The projecting lines for a perspective drawing converge at the eye of the observer and are called *visual rays.* The *picture plane* (PP) is the surface upon which the drawing is made. It is a vertical plane 90 degrees from the horizon plane. Think of it as a piece of glass located between the observer and the object. The *horizon plane* passes through the eye of the observer. The *vanishing points* (VP) are located on the horizon plane. These vanishing points are where all lines converge. Drawings can have more than one vanishing point. The *station point* (SP) represents the position where the observer is standing while viewing the object. The *ground line* (GL) represents the plane upon which the observer is standing. Assuming the observer to be of normal height, the horizon line is located approximately 5'-0'' (eye level) above the ground line. The *measuring line* is a vertical line upon which actual heights are made and from which actual heights are drawn to the vanishing points.

Think of the picture plane (PP) line as a piece of glass set 90 degrees from the ground line (GL). Notice where the observer is standing (SP) and that the horizon line is approximately five feet above the ground line. As you move the picture plane line closer to the object, the object gets larger. The visual rays extend from the observer's eyes to the object.

DRAWING A ONE-POINT PERSPECTIVE

When drawing a one-point perspective, lay out the top view so the proper depth of the object can be located on the lines going to the vanishing points. This avoids guesswork and will result in a more professional looking drawing.

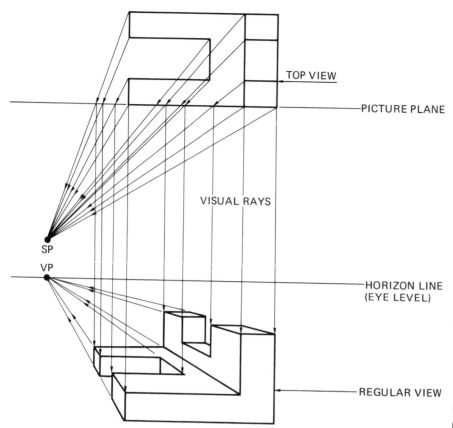

TOP VIEW

PICTURE PLANE

VISUAL RAYS

SP

VP

HORIZON LINE
(EYE LEVEL)

REGULAR VIEW

Fig. 21-13 Method of laying out a one-point perspective

The bottom edge of the top view is placed on the picture plane line, figure 21-13. A point is arbitrarily placed on the paper beneath the picture plane line to represent the station point. This point may be on, to the right, or to the left of the center. It represents where the person viewing the object is standing. The distance this point is away from the picture plane represents the distance the person is away from the object.

A point representing the vanishing point is placed directly beneath the station point any convenient distance apart. A line drawn horizontally through this point represents eye level (horizon line).

Next, the front view of the three-view drawing is drawn directly beneath the top view by projecting lines downward, figure 21-14. The front view may be placed on, above, or below the horizon line depending upon the finished view desired. All edges of the front view are projected to the vanishing point. All edges of the top view not touching the picture plane are drawn to the station point. At the point these projected lines cross the picture plane, they are projected downward to establish the correct depth.

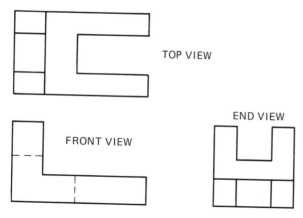

TOP VIEW

END VIEW

FRONT VIEW

Fig. 21-14 Working drawing of the object being drawn in figure 21-13

One-Point Perspectives in Architecture

Figure 21-15 shows one use of one-point perspective in architecture.

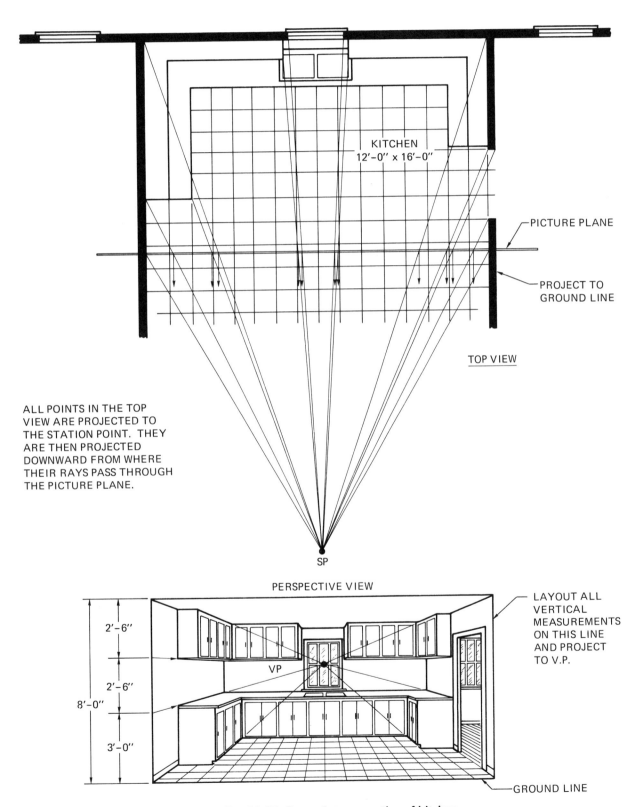

KITCHEN
12'-0'' x 16'-0''

PICTURE PLANE

PROJECT TO
GROUND LINE

TOP VIEW

ALL POINTS IN THE TOP
VIEW ARE PROJECTED TO
THE STATION POINT. THEY
ARE THEN PROJECTED
DOWNWARD FROM WHERE
THEIR RAYS PASS THROUGH
THE PICTURE PLANE.

SP

PERSPECTIVE VIEW

2'-6''

2'-6''

8'-0''

3'-0''

VP

LAYOUT ALL
VERTICAL
MEASUREMENTS
ON THIS LINE
AND PROJECT
TO V.P.

GROUND LINE

Fig. 21-15 One-point perspective of kitchen

DRAWING A TWO-POINT PERSPECTIVE

To begin drawing a two-point perspective, start with the normal, three-view drawing, figure 21-16. Decide which position best illustrates the object.

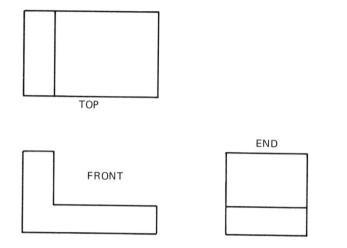

TOP

END

FRONT

Fig. 21-16 Three-view drawing

Sketch the object's basic shape, figure 21-17. Turn the object into the position needed to illustrate all of the important features. Several rough sketches may be needed before making this decision.

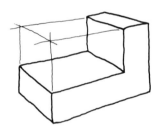

Fig. 21-17 Sketch of object

Select the best sketch and use that layout to make a true two-point perspective as outlined in the following steps.

Step 1

1. Draw a light line horizontally across the paper. This represents the picture plane.
2. Draw the top view at a 30-degree angle to and touching the picture plane, figure 21-18.

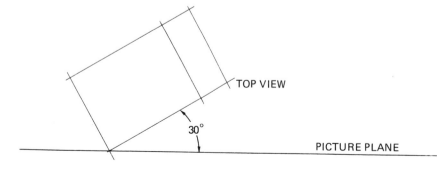

TOP VIEW

30°

PICTURE PLANE

Fig. 21-18 Step 1

Step 2

1. Project a line 90 degrees from the point at which the picture plane and top view touch, figure 21-19.

2. Place a point on the picture plane as far to the right of the paper as possible. Draw line a through this point parallel to the right side of the top view.

3. Station point b is located where line a crosses the line drawn in step 1.

4. Through station point b, draw a line parallel to the left side of the top view.

5. Project lines c and d from the top view to the station point (b). Measure the angle formed by these lines. The angle should be less than 30 degrees. If it is more than 30 degrees either the points on the picture plane or the top view must be moved out. In either case, the entire process must be repeated.

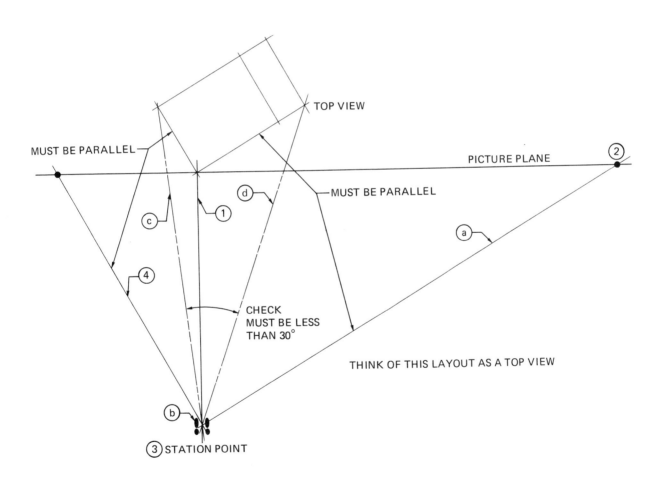

Fig. 21-19 Step 2

Step 3

1. Project lines from each point of the top view downward to the station point. These lines are the visual rays, figure 21-20.
2. Draw the horizon plane representing eye level at some convenient distance below the station point.
3. Project lines downward from points 'A' and 'B' on the picture plane until they intercept the horizon plane. The points at which they intercept become the left vanishing point (LVP) and right vanishing point (RVP).

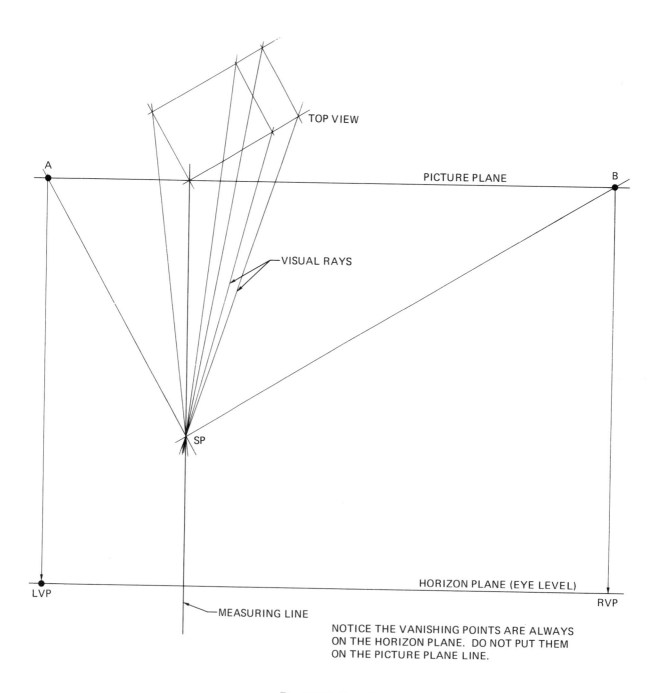

Fig. 21-20 Step 3

Step 4

1. Draw the front view of the object in the position illustrated, placing its base at the ground line. Project true height over to the measuring line, figure 21-21.
2. Project lines downward from the points where the outside visual rays pass through the picture plane.
3. Construct the basic shape of the object by projecting to the vanishing points, as shown.

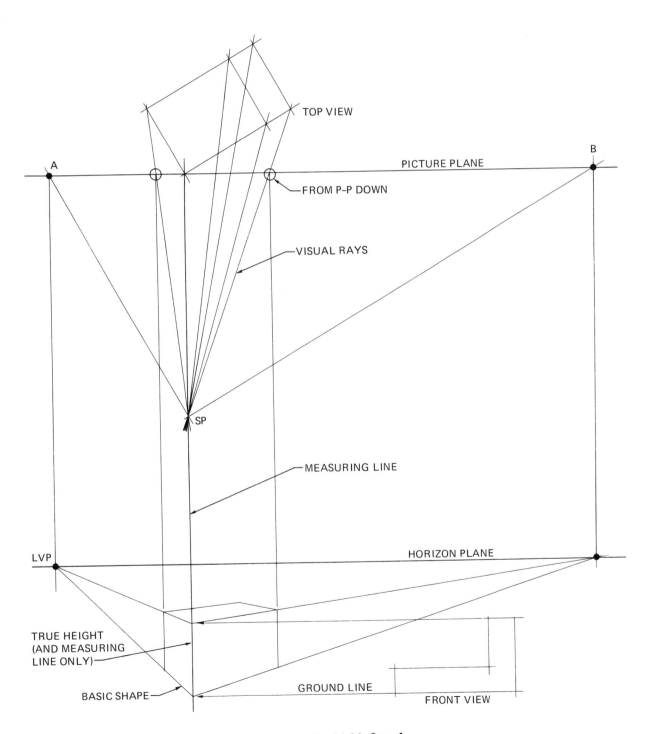

Fig. 21-21 Step 4

Step 5

1. Project downward all other points from the picture plane and complete the
   true perspective view inside the basic shape, figure 21-22.

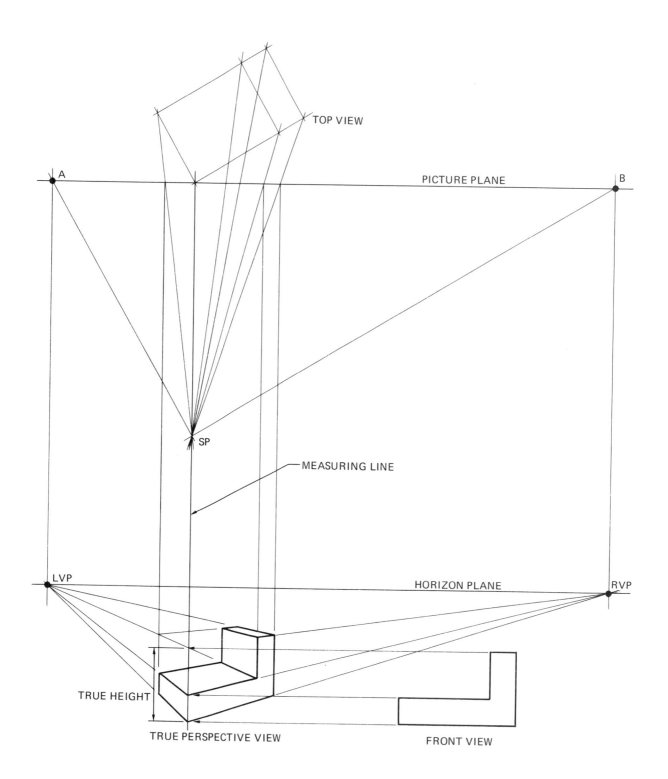

Fig. 21-22 Step 5

A standard three-view drawing is illustrated in figure 21-23, followed by perspective sketches showing the object in several different positions, figure 21-24.

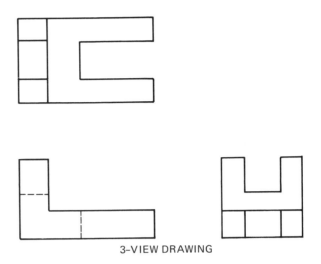

3–VIEW DRAWING

Fig. 21-23 Three-view drawing

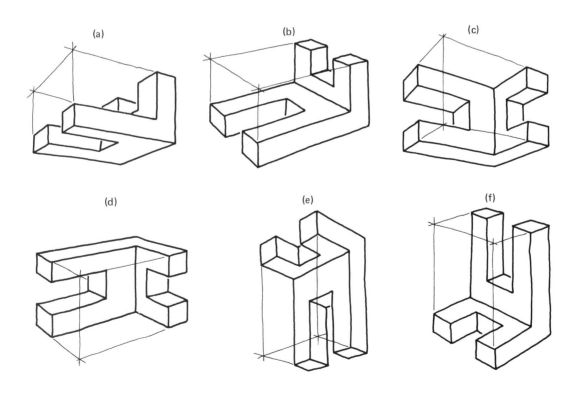

Fig. 21-24 Perspective sketches

SAMPLE PERSPECTIVE LAYOUTS

The elevation selected and the placement of that elevation in relationship to the horizon plane is most important. Study the two examples shown in figure 21-25.

House A is drawn above the horizon plane and appears as if it were located on a hill. House B is drawn below the horizon plane and appears as if it were located in a valley. Neither position is normally used to show a house in perspective. They are used occasionally, however, for special assignments.

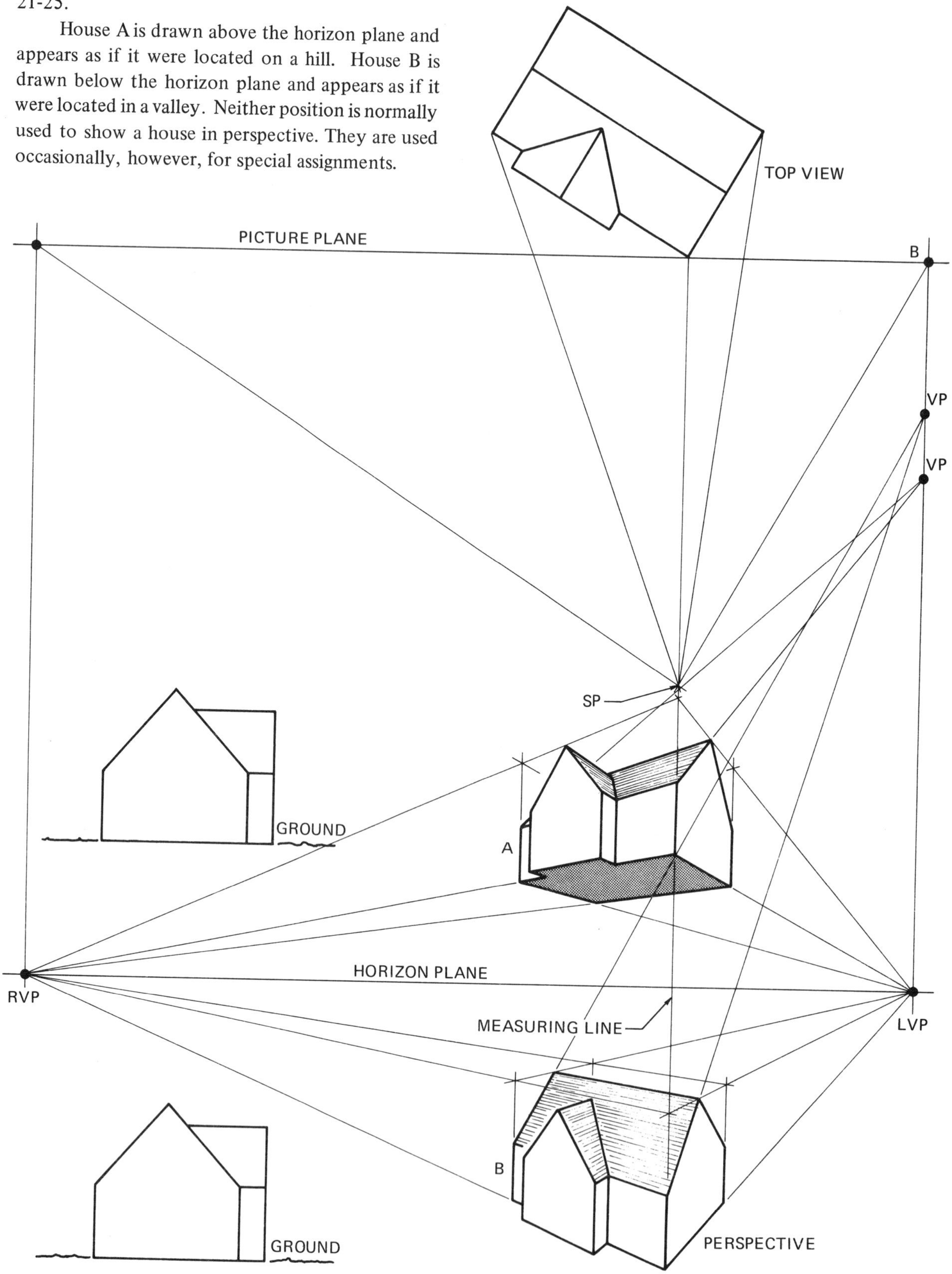

TOP VIEW

PICTURE PLANE

B

VP

VP

SP

GROUND

A

HORIZON PLANE

RVP

MEASURING LINE

LVP

B

GROUND

PERSPECTIVE

Fig. 21-25 Examples of perspective layouts

UNIT 22

THE ENGINEERING DEPARTMENT

DRAFTING DEPARTMENT PRACTICES

All engineering departments operate differently as each varies in size, personnel, function, and scope. New personnel should fully understand the structure of the organization. Engineering organizations must work as efficiently as possible to provide an orderly flow of drawings. To accomplish this, a uniform procedure or standard is followed by all of its members.

An engineering organization provides drawings for new products, improvement of old products, plans for special custom orders, and many other products. This unit explains some of the basic procedures, paper work, and steps necessary to insure the best efficiency. The following items are included:

- Engineering organization
- How to check a drawing
- Parts list (PL)
- Employer-employee agreement
- Typical engineering project flow chart

ORGANIZATION

Figure 22-1 illustrates a small engineering organization employing 26 people, divided into three organizational elements: detailers, clerks, and typists. In larger organizations comprising one or more departments, depending on the work load, drafters are moved from department to department. The higher the position, the more responsibility and the higher the pay.

CHECKING

Though the drafter is responsible for the accuracy of his work, some companies employ a *checker* to double check all drawings. Some of the things a checker looks for include:

- How is the drawing's general appearance? (legibility, neatness, etc.)
- Does it follow all drawing and company standards?
- Are dimensions and instructions clear and understandable?
- Is the drawing easy to understand?
- Are all dimensions included? A machinist must not have to calculate to find a size or location, assume anything, or have any question whatsoever as to what is required.
- Are there unnecessary dimensions?
- Is the drawing prepared so the part may be manufactured the most economical way?
- Will it assemble with mating parts?
- Have all limits, tolerances, and allowances been properly analyzed for all moving parts?
- Have undesirable accumulations of tolerances been adequately analyzed?
- Are all notes added?
- Are finish texture symbols added?
- Is the material and treatment of each part adequate for the design?
- Is the title block complete? Does it include the title, number of the part, drafter's name, and any other required information?

Because of the high cost of errors, it is important that the drafter check and double check work before releasing the drawing. An engineering drawing must be 100 percent correct.

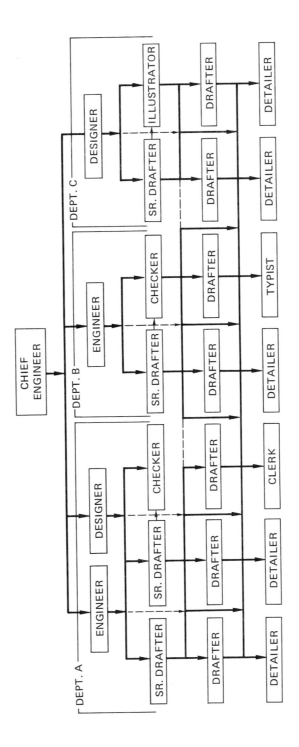

Fig. 22-1 Organization chart

CHANGES

If anyone has a suggestion to improve a part within the product or correct an error on a drawing, it must be brought up at an *engineering change request* (ECR) meeting, figure 22-2. If such suggestions are agreed to by all concerned, an *engineering change order* (ECO) is issued.

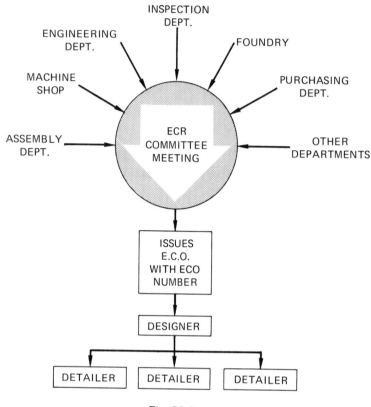

Fig. 22-2

After an engineering change order has been issued, the engineering department assigns a drafter or detailer to make the change(s). After a drawing is issued it cannot be changed by anyone unless it goes through an engineering change request (ECR) and, if approved, an engineering change order (ECO).

In certain cases, administrative changes are appropriate without an engineering change order. Such changes are misspelled words or incorrect references, to improve clarity, etc. These are non-engineering changes. The alternative is to accumulate such changes against the drawing for revision when a more important change is introduced.

The paper work on an engineering change order must include the engineering change request number, why the change was made, who requested the change, what the dimensions were *before* the change, the impact of the change on parts on hand, parts in manufacture, etc.

Change Procedure

After an engineering change order has been issued, the drafter or detailer makes the change.

If the change is extensive: The drawing is completely redrawn and given the same drawing number. The obsolete original is not destroyed but marked "obsolete" and filed.

If the change is fairly simple: The original is carefully erased (so as not to damage the original) where the change is to be made. The change is then carefully added.

If the change is very minor: In the event the change is very minor, such as a small dimension change, the dimension alone is changed and a *wavy line* may be drawn under that dimension indicating it is "out-of-scale." Time is money in an engineering department so all changes must be made in the best and most efficient way. Note that a *change number* is usually placed next to each change and recorded in the *change revision block* on the drawing.

Various companies have different standards or rules governing changes, but the revision block is usually located in a corner of the drawing. It is important that the following is recorded in the revision block.

- Change letter or number (noted beside each change on the drawing also)
- Date of change
- ECR number
- Briefly what was changed
- Name of the person making the change

All changes go through a definite procedure so all departments are notified of a change. Even if the drafter who drew the original finds the error himself after the drawings have been issued, the company's standard change procedure must be followed.

NUMBERING SYSTEMS

All companies have a system of identifying and recording drawings. There is no standard system used by all companies. The most common kind is a letter (A, B, C, D) denoting sheet size, followed by the actual drawing number issued and recorded by a record clerk.

Figure 22-3 shows an example of a detail drawing with four changes (A, B, C, D). Each change was noted by its letter inside a balloon and listed below in the revision record block. The ECO number is also included.

PARTS LIST (PL)

A *parts list* (PL) is a list of all the parts required to assemble a product. It contains the detail number that corresponds to the one on the assembly drawing, the abbreviated drawing title, the plan number, the material, and the quantity required for the assembly. It is the responsibility of the drafter to make up the parts list.

To lay out a parts list:

1. Start with the assembly drawing, first line.
2. List all subassemblies, detail drawings, and purchase parts in the order they are normally assembled, or as you would assemble the unit.
3. Directly after calling out a subassembly, include all detail drawings and purchased parts (indented) that are used to make up that subassembly.
4. Any miscellaneous part (detail drawings, purchased parts) is added at the end of the list.
5. Leave spaces between various subassemblies and parts.

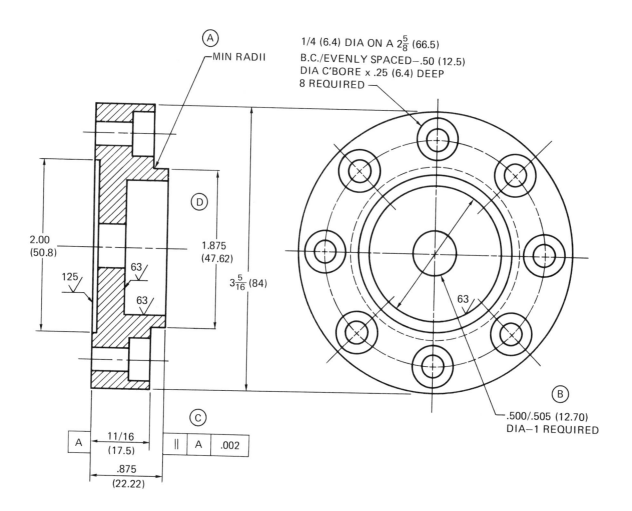

REV.	CHANGE	E.C.O. NO.	BY	CHECKED	DATE
D	WAS 1.870 DIA	776987	PRS	RBC	9 APR 78
C	ADDED TOLERANCE	776953	JAN	RBC	10 JAN 78
B	WAS .375/.380 DIA	776914	CRA	PRS	8 DEC 77
A	ADDED NOTE	776872	JAN	RBC	15 NOV 77
REV.	CHANGE	E.C.O. NO.	BY	CHECKED	DATE
REVISION RECORD					

Fig. 22-3 Recording a change

Parts lists vary from company to company, but most look like the sample in figure 22-4. Carefully study the complete list. Note which parts are sub-assemblies, detail drawings, and purchased parts.

PARTS LIST					
No.	Plan No.	Description		Material	Quan.
1	D-77942	VICE ASSEMBLY – MACHINE		AS NOTED	1
2					
3	C-77947	BASE – VICE		AS NOTED	1
4	B-77952	BASE – LOWER		C.I.	1
5	A-77951	BASE – UPPER		C.I.	1
6	A-77946	SPACER – BASE		STEEL	1
7	PURCH.	BOLT – ½ – 13 UNC – 2″ LG.		STEEL	1
8	PURCH.	NUT – ½ – 13 UNC		STEEL	1
9					
10	C-77955	JAW – SLIDING		STEEL	1
11					
12	A-77954	SCREW – VICE		STEEL	1
13	A-77953	ROD – HANDLE		STEEL	1
14	A-77956	BALL – HANDLE		STEEL	2
15					
16	A-77961	PLATE – JAN		STEEL	2
17	PURCH.	SCREW – ¼ – 20 UNC		STEEL	4
18					
19	A-77962	COLLAR		STEEL	1
20					
21	A-77841	KEY – SPECIAL VICE		STEEL	2
22					
23	PURCH.	BOLT – ½ – 13 UNC – 4″ LG.		STEEL	4
24	PURCH.	NUT – ½ – 13 UNC		STEEL	4

Company Name Company Address	Model No. 999	Parts Lister J.A.N.	Date 5 MAR. 79
Title VICE ASSEMBLY – MACHINE	Page _1_ of _1_ Pages		**DWG No.** D-77942

Fig. 22-4 Parts list

EMPLOYER-EMPLOYEE AGREEMENT

Usually companies that work at designing or creating new devices or processes will have all employees sign an *employer/employee agreement.* Simply stated:

1. You must promise not to reveal any company secrets.
2. If you invent something or discover some new process or method, it belongs to the company.
3. You will not attempt to infringe on any company ideas, processes, or inventions.

This agreement is usually in effect while you are employed and extended for a given time after you leave the company, often six months to two years.

As a rule, you do not get extra compensation for a new design or process, but most companies recognize your efforts and in time adjust your salary or position. In many instances a bonus payment is made in the form of bonds, company stocks, or other forms of recognition.

VARIOUS STANDARDS USED

Companies use various engineering rules or standards. A few such standards include:

- A.N.S.I. – American National Standards Institute
- S.A.E. – Society of Automotive Engineers
- MIL. – Military Standards
- A.S.M.E. – American Society of Mechanical Engineers
- N.E.M.A. – National Electrical Manufacturers Association

Each company also has its own rules or standards to follow.

PERSONAL TECHNICAL FILE

It is important to locate technical information quickly. Usually a company manufacturing a certain product or performing a particular kind of engineering will use certain technical information. A conscientious drafter should develop and update a *personal technical file* containing:

- All company products associated with assignments
- Notes, copies, or clippings from various technical magazines, literature, etc. associated with the company product
- Information on standard materials commonly used in assignments
- Miscellaneous information that would make your job more efficient
- Records of various supervisors, pay levels, and dates of special assignments.
- Various pages of units in this book that were helpful. This material should be neatly organized in a loose leaf binder.

PROJECT FLOW CHART

The process of developing a new product from its established need and conception to its final production consumes many hours and involves highly trained personnel. The flow chart in figure 22-5 gives an example of the process involved.

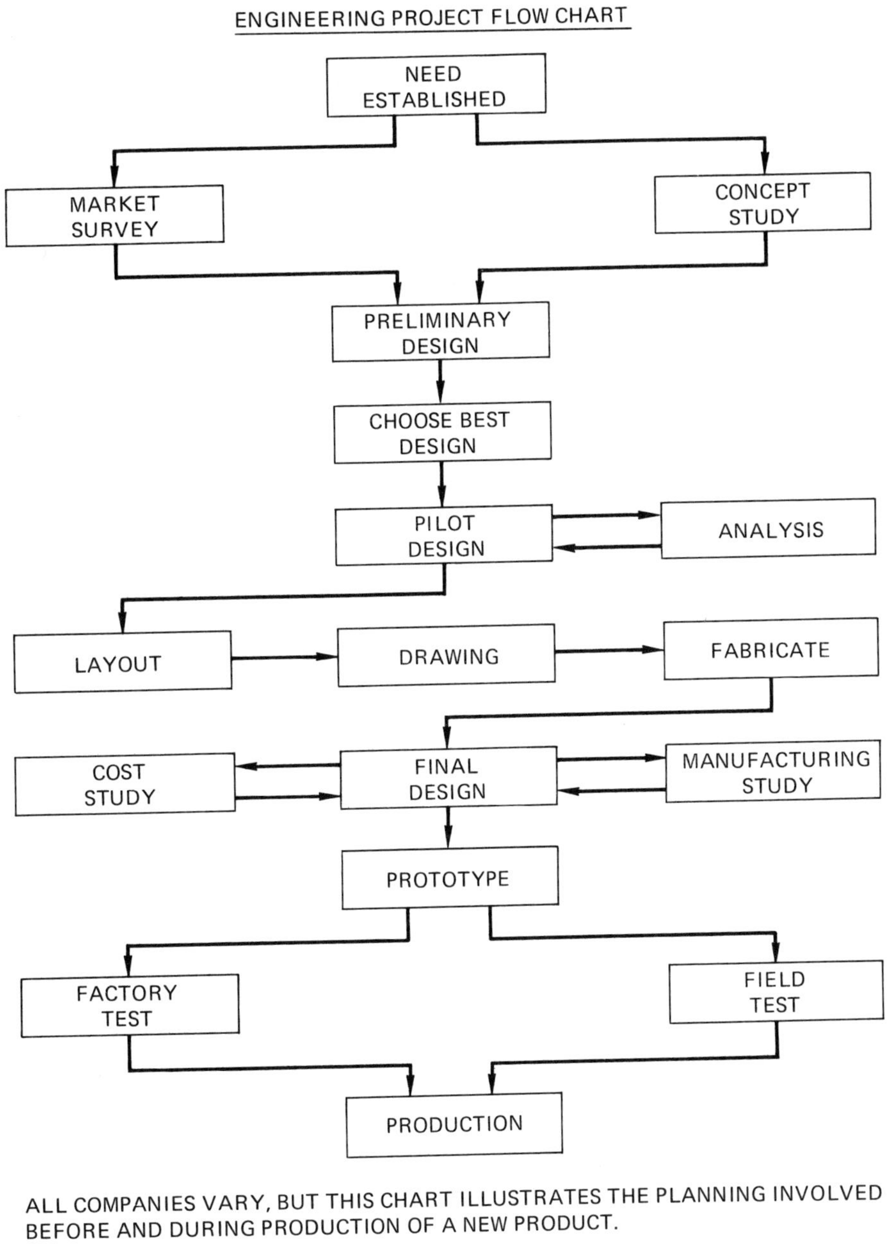

ENGINEERING PROJECT FLOW CHART

ALL COMPANIES VARY, BUT THIS CHART ILLUSTRATES THE PLANNING INVOLVED BEFORE AND DURING PRODUCTION OF A NEW PRODUCT.

Fig. 22-5 Project flow chart

CAREERS IN DRAFTING

The student who develops excellent skills in a particular area of drafting can find many opportunities for employment. Although these positions may be at a junior detailer level, the student could expect to advance to more important positions as knowledge and experience is gained. Each promotion will bring more responsibility and a higher salary. There is no limit to how far the student can advance in drafting as long as the student qualifies for the next step. This requires hard work and study.

The occupational chart in figure 22-6 shows some of the different jobs available in the various fields of drafting.

The next few pages outline various *job descriptions.* Study each one. Notice how a beginning drafter must go through various steps in the drafting field in order to advance in the profession.

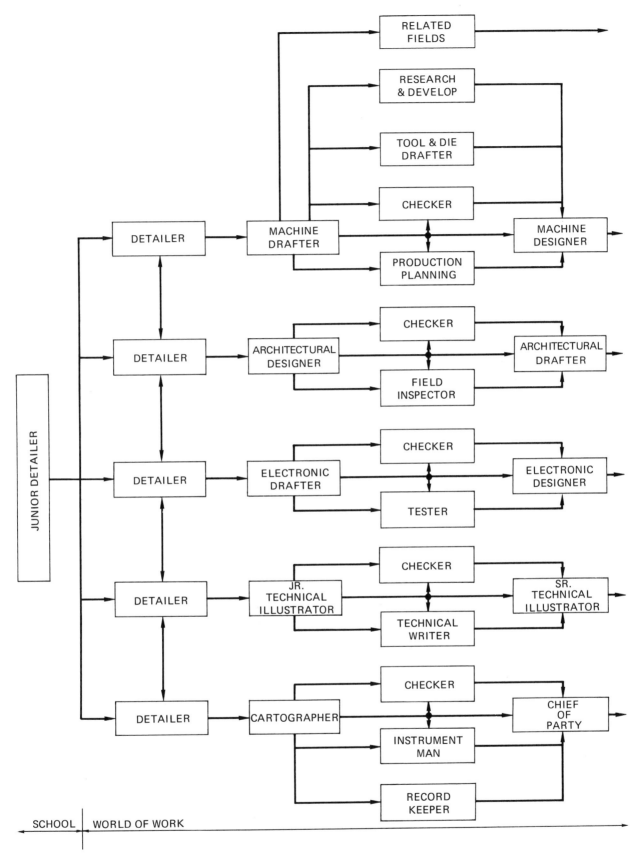

Fig. 22-6 Occupational chart, drafting

TITLE: Junior Detailer

DEPARTMENT: Engineering

REPORTS TO: Chief Drafter (or supervisor)

SUPERVISES: No one

TYPICAL DUTIES:

See detailer. Junior detailer's duties are similar to those of detailer, but limited to less complicated types of drawings requiring less experience.

SPECIFICATIONS:

Education: High school graduate

Experience: None

Job Preparation: This is a beginning job or a promotion from tracer. May be promoted to detailer.

DISTINGUISHING
CHARACTERISTICS:

May be promoted to detailer when able to perform the duties of that classification in a competent manner with a minimum of supervision.

TITLE: Detailer

DEPARTMENT: Engineering

REPORTS TO: Chief Drafter

SUPERVISES: No one

TYPICAL DUTIES:

1. Detailing: Makes detail parts drawings from layouts, minor assembly drawings, setting plans, piping and wiring diagrams, and catalog or repair manual charts and illustrations. Works from sketches, layouts, notes, or current drawings. Makes drawings to scale indicating dimensions, material specifications, and other standard information required to make drawings clear.

2. Changes: Makes changes on current drawings on the basis of change orders received. May prepare change order releases, bills of material, and assist with other clerical work in the engineering department.

SPECIFICATIONS:

Education: High school graduate. Courses in drafting desirable.

Experience: Previous experience as tracer or start directly in this job if adequately trained in drafting skills.

Job Progression: May have been promoted from tracer or started directly in this job if adequately trained in drafting skills. May be promoted to drafter, when qualified.

DISTINGUISHING
CHARACTERISTICS:

Must be able to produce parts and minor assembly drawings which are sufficiently clear, complete, and accurate for use in shop manufacturing operations.

TITLE: Detailer-Drafter

DEPARTMENT: Engineering

REPORTS TO: Chief Drafter

SUPERVISES: No one

TYPICAL DUTIES:

1. Detailing and drafting: Prepares clean and accurate detail and some assembly drawings of some complexity from sketches, notes, current drawings, and verbal instructions. Draws to scale. Establishes dimensions, material specifications, and other information necessary for manufacture. Establishes some mathematical calculations.

2. Changes: Makes some drawing and design changes on the basis of change orders. Makes plan releases, bills of materials, etc.

SPECIFICATIONS:

Education: High school graduate. Courses in drafting desirable.

Experience: Equivalent of two years detailing experience.

Job Progression: An employee in the detailer classification making satisfactory progress may be promoted to this classification after reaching the top rate for detailer. May be promoted sooner, if qualified. An employee adequately trained in detailing may be placed directly in this classification.

DISTINGUISHING
CHARACTERISTICS:

Must be able to prepare accurate finished drawings of parts and assemblies showing dimensions; some mathematical calculations.

TITLE: Drafter

DEPARTMENT: Engineering

REPORTS TO: Chief Drafter

SUPERVISES: No one

TYPICAL DUTIES:

1. Drafting: Prepares clear, complete, and accurate detail and assembly drawings of average complexity from rough sketches, notes, or verbal instructions. Establishes dimensions, machining tolerances, materials, and other information necessary for manufacture. Makes all necessary mathematical calculations.

2. Change Orders: Makes drawing and design changes on the basis of engineering change orders. Writes change orders, lists, and releases.

3. Checking: May do some checking for accuracy of drawings prepared by detailers or tracers.

SPECIFICATIONS:

Education: High school graduate. Courses in drafting desirable.

Experience: Previous experience as a detailer required. May be promoted when qualified and work is available.

Job Progression: Promoted from detailer. May be promoted to layout drafter.

DISTINGUISHING
CHARACTERISTICS:

Must be able to prepare complete and accurate finished drawings of average difficulty showing all dimensions, tolerances, etc., after making all necessary mathematical calculations.

TITLE: Senior Drafter
DEPARTMENT: Engineering
REPORTS TO: Chief Drafter
SUPERVISES: Leads, directs, trains, and checks the work of less experienced drafters.

TYPICAL DUTIES:

1. Drafting: Makes layout drawings of new, revised, special, or salvage parts and assemblies from specifications, drawings, sketches, or general design data. Establishes dimensions. May make complex detail drawings. Makes sketches and original designs from verbal instructions.

2. Analysis: Analyzes layout parts for strength, weight, wear, ease of manufacturing and assembly.

3. Checking: Checks the work of drafters, detailers, and tracers for accuracy and practicality of manufacture.

4. Consultations: Works with the manufacturing and process departments on methods of manufacturing applicable to the products being designed. Checks patterns and sample castings, observes and assists with test operations, and revises drawings according to valid suggestions from other departments.

SPECIFICATIONS:

Education: High school graduate. Appropriate courses in mathematics and drafting.

Experience: No time limit. Must be a thoroughly experienced and versatile drafter, able to do top quality work of the highest complexity without assistance.

Job Progression: Promoted from drafter. May be promoted to checker or design drafter.

DISTINGUISHING
CHARACTERISTICS:

Only drafters who are able to do first class work on the most complex assignments are placed in this classification.

TITLE: Design Drafter

DEPARTMENT: Engineering

REPORTS TO: Chief Engineer

SUPERVISES: Directs, trains, and checks the work of drafters of various grades.

TYPICAL DUTIES:

1. Designing: Designs equipment or components. Makes preliminary investigations, calculations, studies, sketches, and basic working layouts for modifications of current designs or for new products.

2. Calculations: Performs necessary calculations on strength of materials, springs, gears, capacity, working fits, etc.

3. Follow Up: Follows designs through periods of detailing. Checks assemblies and tests. Keeps supervisor advised on difficulties encountered. Consults with personnel in tool design, processing, cost, pattern shop, foundry, etc., on matters of design, manufacture, materials, and cost.

SPECIFICATIONS:

Education: Equivalent of two years college engineering.

Experience: Minimum of five to eight years in drafting, design, and related work.

DISTINGUISHING
CHARACTERISTICS:

Design drafters have a specialized knowledge of a particular product line. They may be as expert as the engineers in this restricted field, but their training is less comprehensive and diversified.

WHERE TO LOOK FOR A JOB

Jobs can be found almost anywhere if a drafter's skills are good. It is often said that the best jobs are not advertised, which means a drafter must devise ways to find them.

Personal Contacts. Make a list of everyone who might know of a job opening or contact. Talk with relatives about their friends and business associates who might serve as references. Do not overlook teachers, clergy, or doctors. Guidance counselors have contacts, booklets, and lists which may be helpful in a job search.

Yellow Pages. Use the Yellow Pages of the phone book to make a list of companies, large and small, who hire drafters. When the list is complete, apply directly to the company or organization. Do not expect an immediate interview. In most cases, an application blank must first be completed.

Trade Associations. A drafter should also investigate *trade associations*. These are organizations of people, companies, and groups in one particular field. Some have their own placement services and most publish their own magazines or newsletters which carry classified advertising about employment opportunities.

Newspapers. Read the newspapers every day. Read every section — the arts, finance, news stories. News items contain valuable information, such as the opening of a new company. Write down the name and address and call them.

Newspaper ads are the most obvious place to look for jobs. Is that opening the right one, or does it just sound good? Sometimes advertising writers make a position sound interesting when it is actually very routine. Do not reply to ads that are not appropriate. Consider such things as the time it takes to travel to the job, how much knowledge and experience the job requires, etc. Follow the ad's instructions exactly. If instructed to "reply in your own handwriting," do so. The job probably requires legible handwriting. Keep a record of queries and replies. Make a carbon of any letter sent out. Clip the ad and staple it to the carbon copy.

Employment Agency. Consider an employment agency. A good one does all the tedious detail work. An agency representative has many more contacts than the average person seeking a job. A great many companies deal only with agencies when hiring new people because they do not have enough time or personnel to screen applicants themselves. An agency can actually carry out a complete job hunt for an applicant. This service is free until a job is secured. At that time a fee, equal to a certain percentage of the monthly wage, is paid to the agency. *Find out what the fee will be before working with an agency.*

After an interview with a company, report back to the agency. The agency may know how the interview was received and what are the chances of getting that particular job. Remember, although a fee may seem large, it is tax deductible. Do not refuse interviews for jobs that do not pay the fee. Take all interviews of interest. Interview experience is invaluable. If a job is offered, however, do not take it unless it is really worth the price of the agency fee.

If job hunting is done with care, more than one job offer may have to be considered. Do not just grab the first one that comes along. Any big decision, like how to spend half a lifetime, requires time to think. Most companies realize this. Even if the right job seems to come along, it is a good policy to ask for time (a few days is reasonable) to think it over.

It is impossible to predict for certain how well a particular job will turn out. If a drafter is conscientious in the search, success is attainable.

WRITING A RESUME

A *resume* is a brief, one-page outline or summary of an applicant's qualifications. A resume should stimulate interest in order to get an interview with a prospective employer. The initial impression from the resume is very important. Take the time to do a thorough self-examination of your attributes before writing it. Be sure to omit any trivial information.

There are many different ways of writing a resume. The important thing to remember is that a resume should be neat, easy to read, and attractively arranged on the page. Figure 22-7 shows an example of a resume.

A resume should contain the following information, though not necessarily in this order:

- Name, address, and telephone number
- Occupational objective
- Work experience
- Education
- Personal data (optional)
- References

Every resume begins with the applicant's full name and address, including zip code, and telephone number.

A resume should include an occupational objective that states exactly what kind of job the applicant is looking for. This should be interesting enough to motivate a prospective employer to read the rest of the resume.

Work experience should include the names and addresses of employers, job title, job description, dates of employment, and supervisor's name. List jobs in reverse chronological order, starting with the most recent job and working backwards. Try to highlight strong points and important accomplishments.

Educational information includes the name and address of all institutions attended, special training, whether or not the applicant graduated, and the degree or course of study. If it is pertinent to the job or highlights an impressive quality, mention such things as class standing, honors, extracurricular activities, etc. Dates of graduation and attendance need not be listed.

Resume of

JAMES A. BROWN
107 Elmwood Drive
Palma, CA 04402

CAREER OBJECTIVE

To enter the drafting field as a detailer in order to gain basic experience and qualify, by performance, to become a drafter.

EDUCATION

Hamilton High School; 17 Oak Street; Palma, CA 04402

Graduated: June 1979
Major: Vocational drafting
Class Standing: 15th in a class of 87
Honors: Architectural Drafting Award, Spring 1979
Activities: VICA, 1–4; Baseball 3–4

Dixon County Community College; Adult Education Program; Beacon, CA 05904

Summer 1979
Course: Landscaping (ten week course)

WORK EXPERIENCE

June, 1977 to present   Eagle Supermarket; 1550 West Drive; Beacon, CA 05904

Supervisor: Mr. Albert Smith
Position: Clerk (part-time). Stocked shelves and displays, cashier, deliveries

Summer, 1978   Palma Summer Theatre; 200 Sunset Road, Palma, CA 04402

Director: Mrs. Lillian Garver
Position: I worked on the stage crew constructing sets for the theatre's five summer stock productions.

PERSONAL DATA

Birth date: April 14, 1961
Marital Status: Single
Social Security: #005-44-0045
Health: Good

REFERENCES

Mr. Robert Eckerle, Set Designer; Palma Summer Theatre; 200 Sunset Road; Palma, CA 04402
The Rev. Roger Fredericks; First Lutheran Church; 12 Church Street; Palma, CA 04402
Miss Jennifer May, Drafting Teacher; Hamilton High School; 17 Oak Street; Palma, CA 04402

Fig. 22-7 Sample resume

If it is to the applicant's advantage, personal data may be included on the resume. However, the applicant does not have to provide information regarding age, marital status, sex, number of dependents, race, creed, physical limitations, etc. The applicant may wish to do so if it will increase the chance of securing a job. Outside activities and hobbies, for instance, may indicate areas of competence that will impress an employer.

References should also be included on the resume. If there is not enough room, simply state that "references are available upon request" at the bottom of the resume. These references should include two people who can attest to the applicant's occupational skills and willingness to do a good job. A third reference should be able to vouch for the applicant's character.

Do not include the date on a resume. This is written on the cover letter. Do not include salary requirements. This is discussed during the personal interview. Do not forget that the resume forms an employer's first opinion of an applicant. A resume that rambles and is confusing will find its way into the wastebasket. Be sure the resume is neat, clearly typed, and includes all necessary information.

THE COVER LETTER

A *cover letter* accompanies a resume. It is a short, concise letter designed to arouse a prospective employer's interest in the candidate. It should convince the employer to read the resume and set up an interview. Figure 22-8 shows a sample cover letter.

The opening paragraph of a cover letter states the purpose of the letter — that the applicant feels he or she is the best possible choice for the job. The first paragraph also mentions how the applicant found out about the job or whether this is a letter of inquiry about any available positions.

The middle paragraph(s) explains why the candidate feels he or she is the right person for the job. This is a good place to stress qualities that may not be immediately apparent by reading the resume, such as an applicant's ability to work with others or to handle responsibility. Remember the cover letter does not simply repeat what is in the resume, it explains and expands the data. After reading the cover letter, the employer will *want* to read the resume to learn more about the applicant.

The closing paragraph usually asks for an interview. Applicants should state that they are available for an interview at the company's convenience and how they can be reached (telephone number, etc.).

107 Elmwood Drive
Palma, CA 04402
September 23, 1979

Mr. Edwin A. Larson, Personnel Director
North-Central Home Manufacturing, Inc.
395 Colvin Avenue
Los Angeles, CA 90014

Dear Mr. Larson:

I feel my high school training in drafting qualifies me for the position of detailer advertised in the Los Angeles Tribune on September 22, 1979.

As indicated on the enclosed resume, I recently graduated from Hamilton High School where I participated in a comprehensive two-year drafting program. This included an intensive one year drawing course in basic drafting skills. The second year concentrated on architectural, technical illustration, civil, and mechanical drafting. I received the highest average in architectural drafting. My drawings of a contemporary residence were displayed at the Palma First National Bank during Student Appreciation Week, June 12–17, 1979.

To supplement my high school studies, I took a summer course in landscaping to learn how plants can improve exterior house design. In addition, by working with the stage crew at Palma Summer Theatre, I gained practical experience in construction. I enjoyed working with others towards a common goal in both of these ventures.

I am available for an interview at your convenience and can be reached at 555-2000 every weekday morning. At that time, I can bring a few of the drawings I made in the architectural drafting course.

Sincerely,

James A. Brown

James A. Brown

Fig. 22-8 Sample cover letter

THE INTERVIEW

If the cover letter and resume show that the applicant has the qualifications for the job, a company representative will most likely invite the applicant to a personal interview. During the interview, the prospective employer will seek information about the applicant that is not available on the application or resume. In return, the applicant asks about salary, future progression, benefits, etc. The interview, therefore, benefits both the employer and the applicant.

An interview usually lasts twenty or thirty minutes. Its main purpose is to give the applicant a chance to tell why he or she is the right person for the job. Some questions which may be asked during the interview include:

- Why do you want to be a drafter?
- What qualifies you to be a drafter?
- Why did you pick this company?
- Which position interests you most?
- What other jobs have you had?
- Can you take and follow instructions?
- What do you expect to be doing in ten years?
- What pay scale do you expect to start at?
- What have you done that shows initiative?
- Do you have any special abilities?

Confidence impresses an employer. An applicant should give a relaxed and self-controlled appearance. It helps to dress neatly and to arrive 15 minutes early. Greet the interviewer with a firm handshake and maintain eye contact throughout the interview. Know as much about the company as possible before the interview and decide which questions to ask during the interview. Be sincere, poised, and in control. Applicants should also bring a few drawing samples to illustrate their level of competence.

During the interview, the interviewer will often fill out an evaluation form similar to the one in figure 22-9.

Upon returning home from an interview, the applicant should write a short letter thanking the interviewer for the meeting. This letter stresses that the interview was informative and that the applicant still feels he or she is right for the job and company. An example of a follow-up letter appears in figure 22-10.

NAME __James A. Brown__   PHONE __716-3257__

POSITION APPLIED FOR __Detailer__

DRESS			
Careless	Neat	Very neat	**5**
1 2	3 4 ⑤	6	

APPEARANCE-HEALTH			
Sickly	Good health	Excellent health	**5**
1 2	3 4 ⑤	6	

IS APPLICANT AT EASE			
Embarrassed	Not at ease	At ease	**4**
1 2	3 ④ 5	6	

VOICE			
Unpleasant	Pleasant	Very pleasant	**5**
1 2	3 4 ⑤	6	

TRAINING FOR JOB (EDUCATION - EXPERIENCE)			
Lacks training	Good training/fair experience	Excellent	**3**
1 2	③ 4	5 6	

ATTITUDE			
Over-bearing	Reasonable	Excellent	**5**
1 2	3 4 ⑤	6	

PERSONALITY			
Unstable	Satisfactory	Outstanding	**5**
1 2	3 4 ⑤	6	

REMARKS __Direct from H.S.__

__Good education / No work exper.__

TOTAL RATING	**32**

Fig. 22-9 Sample evaluation form

107 Elmwood Drive
Palma, CA 04402
September 30, 1979

Mr. Edwin A. Larson, Personnel Director
North-Central Home Manufacturing, Inc.
395 Colvin Avenue
Los Angeles, CA 90014

Dear Mr. Larson:

I certainly enjoyed our interview on Wednesday and appreciate the time you spent telling me about North-Central Home Manufacturing and the position of detailer.

Judging from your description of the position, I feel I can meet the company's rigid specifications. The potential for learning and advancement within the company convinces me even further that I would like to work at North-Central.

Thank you for this opportunity to discuss my qualifications.

Sincerely,

James A. Brown

James A. Brown

Fig. 22-10 Sample follow-up thank you letter

EMPLOYMENT APPLICATION

For proper consideration answer all questions completely & accurately - Print in black ink

Type of work applied for

Last name	First	Middle	Social Security Number

Home address (no. street & State

Phone number

U.S. Citizen	☐ Yes ☐ No	Are you between the ages of 18 and 65?	☐ Yes	☐ No

Do you have any emotional or physical limitations that will prevent you from performing the job you are applying for? ☐ Yes   ☐ No

Have you ever been convicted of a crime? If yes, explain.

School	Location	Did You Graduate?	Major	Degree
H.S.				
TECH				
COLLEGE				
OTHER				

List any special training

List all full time, part-time jobs held			
Name/Address	Type of work	Date	Reason for leaving

References (3)		
Name	Address	Occupation

Signature _____

Fig. 22-11 Sample application form

THE APPLICATION FORM

At some point before the interview, all applicants must fill out an *application form,* figure 22-11. These are usually from two to four pages long. Read all instructions before filling in an application. Some forms must be lettered in black ink or typed.

Lightly fill out the application first. Check all answers, particularly dates, names, addresses, telephone numbers, and spelling. If a question does not apply, place a dash through it so the interviewer knows it was read but does not apply. Once everything has been checked, complete the application in ink or as directed. Be neat. Remember, that as a drafter, legible lettering is an important skill.

INVENTION AGREEMENT

Anyone who works in a job that is creative, such as drafters and engineers, must sign an agreement form giving the company the right to any new invention designed while working for the company. The form usually puts in writing that the employee will not reveal any of the company's discoveries or projects.

The invention agreement is in effect for six months to two years after an employee leaves the company, depending upon the company. This is so employees will not invent something, quit, and patent it themselves.

A company does not usually give extra pay for an invention. This is what an employee is paid to do. However, a company usually recognizes talent and will reward an employee with promotions, stocks, bonds, or possibly a raise.

Signing an employer/employee invention agreement is a normal request in any company. Read the contract before signing it. However, if an applicant does not sign the agreement, he or she will probably not be offered the job.

A PROFESSIONAL ATTITUDE

Being a skilled and efficient drafter is only part of the job. Drafters must always conduct themselves in a professional manner as well. This professional attitude is not easy to define, but here are some of the characteristics associated with true professionalism.

Professionals:

- Do not require close supervision or direction. They plan their own activities and work independently.

- Regard their supervisors as fellow professional workers and, in return, are treated the same way.

- Adjust their working hours to meet the necessities and responsibilities of the job, even when this requires working overtime.

- Take full responsibility for the results of their efforts and actions. They seek advice and counsel but do not attempt to transfer responsibility for their own mistakes to others.

- Continually seek self-improvement and take advantage of every opportunity to learn.

- Contribute to the skill and knowledge of their profession by developing new ideas, plans, and materials which are gladly shared with fellow workers.

- Respect the confidence of others. The welfare of others often requires that information concerning them remain confidential.

- Are loyal to fellow workers and to those they serve.

- Avoid rumor and hearsay. Professionals secure information only from those authorized to release it.

- Meet their professional obligations. The professional completes all agreements, whether legal or moral obligations.

- Do not advance themselves at the expense of others. Professionals strive for promotion only through their own performance.

- Are proud of their profession. They always reflect this pride and satisfaction with their work to those outside their profession.

APPENDIX A

	INCH/METRIC – EQUIVALENTS					
	Decimal Equivalent				**Decimal Equivalent**	
Fraction	Customary (in.)	Metric (mm)		**Fraction**	Customary (in.)	Metric (mm)
1/64 — .015625		0.3969		33/64 — .515625		13.0969
1/32 — .03125		0.7938		17/32 — .53125		13.4938
3/64 — .046875		1.1906		35/64 — .546875		13.8906
1/16 — .0625		1.5875		9/16 — .5625		14.2875
5/64 — .078125		1.9844		37/64 — .578125		14.6844
3/32 — .09375		2.3813		19/32 — .59375		15.0813
7/64 — .109375		2.7781		39/64 — .609375		15.4781
1/8 — .1250		3.1750		5/8 — .6250		15.8750
9/64 — .140625		3.5719		41/64 — .640625		16.2719
5/32 — .15625		3.9688		21/32 — .65625		16.6688
11/64 — .171875		4.3656		43/64 — .671875		17.0656
3/16 — .1875		4.7625		11/16 — .6875		17.4625
13/64 — .203125		5.1594		45/64 — .703125		17.8594
7/32 — .21875		5.5563		23/32 — .71875		18.2563
15/64 — .234375		5.9531		47/64 — .734375		18.6531
1/4 — .250		6.3500		3/4 — .750		19.0500
17/64 — .265625		6.7469		49/64 — .765625		19.4469
9/32 — .28125		7.1438		25/32 — .78125		19.8438
19/64 — .296875		7.5406		51/64 — .796875		20.2406
5/16 — .3125		7.9375		13/16 — .8125		20.6375
21/64 — .328125		8.3384		53/64 — .828125		21.0344
11/32 — .34375		8.7313		27/32 — .84375		21.4313
23/64 — .359375		9.1281		55/64 — .859375		21.8281
3/8 — .3750		9.5250		7/8 — .8750		22.2250
25/64 — .390625		9.9219		57/64 — .890625		22.6219
13/32 — .40625		10.3188		29/32 — .90625		23.0188
27/64 — .421875		10.7156		59/64 — .921875		23.4156
7/16 — .4375		11.1125		15/16 — .9375		23.8125
29/64 — .453125		11.5094		61/64 — .953125		24.2094
15/32 — .46875		11.9063		31/32 — .96875		24.6063
31/64 — .484375		12.3031		63/64 — .984375		25.0031
1/2 — .500		12.7000		1 — 1.000		25.4000

APPENDIX B

CIRCUMFERENCES AND AREAS (0.2 to 9.8; 10 to 99)*

Diameter	Circum.	Area	Diameter	Circum.	Area	Diameter	Circum.	Area
0.2	0.628	0.0314	11	34.55	95.03	56	175.9	2,463
0.4	1.26	0.1256	12	37.69	113	57	179.1	2,551.8
0.6	1.88	0.2827	13	40.84	132.7	58	182.2	2,642.1
0.8	2.51	0.5026	14	43.98	153.9	59	185.4	2,734
1	3.14	0.7854	15	47.12	176.7	60	188.5	2,827.4
1.2	3.77	1.131	16	50.26	201	61	191.6	2,922.5
1.4	4.39	1.539	17	53.4	226.9	62	194.8	3,019.1
1.6	5.02	2.011	18	56.54	254.4	63	197.9	3,117.3
1.8	5.65	2.545	19	59.69	283.5	64	201.1	3,217
2	6.28	3.142	20	62.83	314.1	65	204.2	3,318.3
2.2	6.91	3.801	21	65.97	346.3	66	207.3	3,421.2
2.4	7.53	4.524	22	69.11	380.1	67	210.5	3,525.7
2.6	8.16	5.309	23	72.25	415.4	68	213.6	3,631.7
2.8	8.79	6.158	24	75.39	452.3	69	216.8	3,739.3
3	9.42	7.069	25	78.54	490.8	70	219.9	3,848.5
3.2	10.05	7.548	26	81.68	530.9	71	223.1	3,959.2
3.4	10.68	8.553	27	84.82	572.5	72	226.2	4,071.5
3.6	11.3	10.18	28	87.96	615.7	73	229.3	4,185.4
3.8	11.93	11.34	29	91.1	660.5	74	232.5	4,300.8
4	12.57	12.57	30	94.24	706.8	75	235.6	4,417.9
4.2	13.19	13.85	31	97.39	754.8	76	238.8	4,536.5
4.4	13.82	15.21	32	100.5	804.2	77	241.9	4,656.6
4.6	14.45	16.62	33	103.7	855.3	78	245	4,778.4
4.8	15.08	18.1	34	106.8	907.9	79	248.2	4,901.7
5	15.7	19.63	35	110	962.1	80	251.3	5,026.6
5.2	16.33	21.24	36	113.1	1,017.9	81	254.5	5,153
5.4	16.96	22.9	37	116.2	1,075.2	82	257.6	5,281
5.6	17.59	24.63	38	119.4	1,134.1	83	260.8	5,410.6
5.8	18.22	26.42	39	122.5	1,194.6	84	263.9	5,541.8
6	18.84	28.27	40	125.7	1,256.6	85	267.0	5,674.5
6.2	19.47	30.19	41	128.8	1,320.3	86	270.2	5,808.8
6.4	20.1	32.17	42	131.9	1,385.4	87	273.3	5,944.7
6.6	20.73	34.21	43	135.1	1,452.2	88	276.5	6,082.1
6.8	21.36	36.32	44	138.2	1,520.5	89	279.6	6,221.2
7	21.99	38.48	45	141.4	1,590.4	90	282.7	6,361.7
7.2	22.61	40.72	46	144.5	1,661.9	91	285.9	6,503.9
7.4	23.24	43.01	47	147.7	1,734.9	92	289.0	6,647.6
7.6	23.87	45.36	48	150.8	1,809.6	93	292.2	6,792.9
7.8	24.5	47.78	49	153.9	1,885.7	94	295.2	6,939.8
8	25.13	50.27	50	157.1	1,963.5	95	298.5	7,088.2
8.2	25.76	52.81	51	160.2	2,042.8	96	301.6	7,238.2
8.4	26.38	55.42	52	163.4	2,123.7	97	304.7	7,389.8
8.6	27.01	58.09	53	166.5	2,206.2	98	307.9	7,543.0
8.8	27.64	60.82	54	169.6	2,290.2	99	311.9	7,697.7
9	28.27	63.62	55	172.8	2,375.8			
9.2	28.9	66.48						
9.4	29.53	69.4						
9.6	30.15	72.38						
9.8	30.78	75.43						
10	31.41	78.54						

*The formulas for circumference and area of circles are the same regardless of the system of measurement, so these values are accurate for both inches and millimetres.

BEND ALLOWANCE FOR 90° BENDS (INCH)

Radii / Thickness	.031	.063	.094	.125	.156	.188	.219	.250	.281	.313	.344	.375	.438	.500	.531	.625
.013	.058	.108	.157	.205	.254	.304	.353	.402	.450	.501	.549	.598	.697	.794	.843	.991
.016	.060	.110	.159	.208	.256	.307	.355	.404	.453	.503	.552	.600	.699	.796	.845	.993
.020	.062	.113	.161	.210	.259	.309	.358	.406	.455	.505	.554	.603	.702	.799	.848	.995
.022	.064	.114	.163	.212	.260	.311	.359	.408	.457	.507	.556	.604	.703	.801	.849	.997
.025	.066	.116	.165	.214	.263	.313	.362	.410	.459	.509	.558	.607	.705	.803	.851	.999
.028	.068	.119	.167	.216	.265	.315	.364	.412	.461	.511	.560	.609	.708	.805	.854	1.001
.032	.071	.121	.170	.218	.267	.317	.366	.415	.463	.514	.562	.611	.710	.807	.856	1.004
.038	.075	.126	.174	.223	.272	.322	.371	.419	.468	.518	.567	.616	.715	.812	.861	1.008
.040	.077	.127	.176	.224	.273	.323	.372	.421	.469	.520	.568	.617	.716	.813	.862	1.010
.050		.134	.183	.232	.280	.331	.379	.428	.477	.527	.576	.624	.723	.821	.869	1.017
.064		.144	.192	.241	.290	.340	.389	.437	.486	.536	.585	.634	.732	.830	.878	1.026
.072			.198	.247	.296	.346	.394	.443	.492	.542	.591	.639	.738	.836	.885	1.032
.078			.202	.251	.300	.350	.399	.447	.496	.546	.595	.644	.743	.840	.889	1.036
.081			.204	.253	.302	.352	.401	.449	.498	.548	.598	.646	.745	.842	.891	1.038
.091			.212	.260	.309	.359	.408	.456	.505	.555	.604	.653	.752	.849	.898	1.045
.094			.214	.262	.311	.361	.410	.459	.507	.558	.606	.655	.754	.851	.900	1.048
.102				.268	.317	.367	.416	.464	.513	.563	.612	.661	.760	.857	.906	1.053
.109				.273	.321	.372	.420	.469	.518	.568	.617	.665	.764	.862	.910	1.058
.125				.284	.333	.383	.432	.480	.529	.579	.628	.677	.776	.873	.922	1.069
.156					.355	.405	.453	.502	.551	.601	.650	.698	.797	.895	.943	1.091
.188						.427	.476	.525	.573	.624	.672	.721	.820	.917	.966	1.114
.203								.535	.584	.634	.683	.731	.830	.928	.976	1.124
.218								.546	.594	.645	.693	.742	.841	.938	.987	1.135
.234								.557	.606	.656	.705	.753	.852	.950	.998	1.146
.250								.568	.617	.667	.716	.764	.863	.961	1.009	1.157

EXAMPLE: MATERIAL THICKNESS = 1/8'', INSIDE RADII = 1/4'' R. WHERE THESE CROSS = .284'' BEND ALLOWANCE (B/A) PLUS TOTAL OF STRAIGHT LENGTHS = DEVELOPED LENGTH.

BEND ALLOWANCE FOR 90° BENDS (MILLIMETRE)

Radii / Thickness	0.80	1.58	2.38	3.18	3.96	4.76	5.56	6.35	7.15	7.94	8.74	9.52	11.12	12.70	13.50	15.85
.330	1.46	2.74	3.98	5.22	6.45	7.73	8.96	10.20	11.44	12.71	13.95	15.19	17.70	20.18	21.42	25.16
.406	1.52	2.80	4.04	5.27	6.51	7.78	9.02	10.26	11.50	12.77	14.01	15.24	17.76	20.23	21.47	25.22
.508	1.59	2.87	4.11	5.34	6.58	7.86	9.09	10.33	11.57	12.84	14.08	15.32	17.83	20.30	21.54	25.29
.559	1.63	2.91	4.14	5.38	6.62	7.89	9.13	10.37	11.60	12.88	14.12	15.35	17.86	20.34	21.57	25.32
.635	1.68	2.96	4.20	5.43	6.67	7.95	9.18	10.42	11.66	12.93	14.17	15.40	17.92	20.39	21.63	25.38
.711	1.74	3.02	4.25	5.49	6.72	8.00	9.24	10.47	11.71	12.99	14.22	15.46	17.98	20.44	21.68	25.43
.813	1.81	3.08	4.32	5.56	6.79	8.07	9.30	10.54	11.78	13.05	14.29	15.53	18.04	20.52	21.75	25.50
.965	1.91	3.19	4.42	5.66	6.90	8.18	9.41	10.65	11.89	13.16	14.40	15.64	18.15	20.62	21.86	25.61
1.016	1.95	2.23	4.46	5.70	6.94	8.21	9.45	10.69	11.92	13.20	14.44	15.67	18.19	20.66	21.90	25.64
1.270		3.40	4.64	5.88	7.11	8.39	9.63	10.86	12.10	13.38	14.61	15.85	18.36	20.84	22.07	25.82
1.625		3.65	4.89	6.13	7.36	8.64	9.88	11.11	12.35	13.63	14.86	16.10	18.61	21.08	22.32	26.07
1.829			5.03	6.27	7.51	8.78	10.02	11.26	12.49	13.77	15.01	16.24	18.76	21.23	22.47	26.22
1.981			5.14	6.38	7.61	8.89	10.13	11.36	12.60	13.88	15.11	16.35	18.86	21.34	22.57	26.32
2.058			5.19	6.43	7.67	8.94	10.18	11.42	12.65	13.93	15.17	16.40	18.92	21.39	22.63	26.38
2.311			5.37	6.60	7.85	9.12	10.36	11.60	12.83	14.11	15.35	16.58	19.09	21.57	22.80	26.55
2.388			5.43	6.66	7.90	9.18	10.41	11.65	12.89	14.16	15.40	16.64	19.15	21.62	22.86	26.61
2.591				6.81	8.04	9.31	10.55	11.79	13.03	14.30	15.54	16.78	19.29	21.76	23.00	26.75
2.769				6.93	8.17	9.44	10.68	11.92	13.15	14.43	15.67	16.90	19.42	21.89	23.13	26.88
3.175				7.22	8.45	9.73	10.96	12.20	13.44	14.71	15.95	17.19	19.70	22.17	23.41	27.16
3.962					9.00	10.28	11.52	12.75	13.99	15.27	16.74	17.74	20.25	22.72	23.96	27.71
4.775					9.58	10.85	12.09	13.32	14.56	15.84	17.07	18.31	20.82	23.30	24.53	28.28
5.156						11.12	12.36	13.59	14.83	16.11	17.34	18.58	21.09	23.57	24.80	28.55
5.537								13.85	15.10	16.37	17.61	18.85	21.36	23.83	25.07	28.82
5.944								14.15	15.38	16.66	17.89	19.13	21.64	24.12	25.35	29.10
6.350								14.43	15.67	16.94	18.18	19.42	21.93	24.40	25.64	29.39

EXAMPLE: MATERIAL THICKNESS = 3.18 MM, INSIDE RADII = 6.35 MM R. WHERE THESE CROSS = 12.20 MM BEND ALLOWANCE (B/A) PLUS TOTAL OF STRAIGHT LENGTHS = DEVELOPED LENGTH.

APPENDIX C-3

BEND ALLOWANCE FOR EACH 1° OF BEND (INCH)

Radii / Thickness	.031	.063	.094	.125	.156	.188	.219	.250	.281	.313	.344	.375	.438	.500	.531	.625
.013	.00064	.00120	.00174	.00228	.00282	.00338	.00392	.00446	.00500	.00556	.00610	.00664	.00774	.00883	.00937	.01101
.016	.00067	.00122	.00176	.00231	.00285	.00342	.00395	.00449	.00503	.00559	.00613	.00667	.00777	.00885	.00939	.01103
.020	.00069	.00125	.00179	.00233	.00287	.00343	.00397	.00452	.00506	.00561	.00616	.00670	.00780	.00888	.00942	.01106
.022	.00071	.00127	.00181	.00235	.00289	.00345	.00399	.00453	.00508	.00563	.00617	.00672	.00782	.00890	.00944	.01108
.025	.00074	.00129	.00184	.00238	.00292	.00348	.00402	.00456	.00510	.00566	.00610	.00674	.00784	.00892	.00946	.01110
.028	.00076	.00132	.00186	.00240	.00294	.00350	.00404	.00458	.00512	.00568	.00622	.00676	.00786	.00894	.00948	.01112
.032	.00079	.00134	.00189	.00243	.00297	.00353	.00407	.00461	.00515	.00571	.00625	.00679	.00789	.00897	.00951	.01115
.038	.00084	.00140	.00194	.00248	.00302	.00358	.00412	.00466	.00520	.00576	.00630	.00684	.00794	.00902	.00946	.01120
.040	.00085	.00141	.00195	.00249	.00303	.00359	.00413	.00468	.00522	.00577	.00632	.00686	.00796	.00904	.00958	.01122
.050		.00149	.00203	.00258	.00312	.00368	.00422	.00476	.00530	.00586	.00640	.00694	.00804	.00912	.00966	.01130
.064		.00160	.00214	.00268	.00322	.00378	.00432	.00486	.00540	.00596	.00650	.00704	.00814	.00922	.00976	.01140
.072			.00220	.00274	.00328	.00384	.00438	.00492	.00546	.00602	.00656	.00710	.00820	.00929	.00983	.01147
.078			.00225	.00279	.00333	.00389	.00443	.00497	.00551	.00607	.00661	.00715	.00825	.00933	.00987	.01152
.081			.00227	.00281	.00335	.00391	.00445	.00499	.00554	.00609	.00664	.00718	.00828	.00936	.00990	.01154
.091			.00235	.00289	.00343	.00399	.00453	.00507	.00561	.00617	.00671	.00725	.00835	.00944	.00998	.01162
.094			.00237	.00291	.00346	.00401	.00456	.00510	.00564	.00620	.00674	.00728	.00838	.00946	.00999	.01164
.102				.00298	.00352	.00408	.00462	.00516	.00570	.00626	.00680	.00734	.00844	.00952	.01006	.01170
.109				.00303	.00357	.00413	.00467	.00521	.00575	.00631	.00685	.00739	.00849	.00958	.01012	.01176
.125				.00316	.00370	.00426	.00480	.00534	.00588	.00644	.00698	.00752	.00862	.00970	.01024	.01188
.156					.00394	.00450	.00504	.00558	.00612	.00668	.00722	.00776	.00886	.00994	.01048	.01212
.188						.00475	.00529	.00583	.00637	.00693	.00747	.00802	.00911	.01019	.01073	.01237
.203								.00595	.00649	.00704	.00759	.00813	.00923	.01031	.01085	.01249
.218								.00606	.00660	.00716	.00770	.00824	.00934	.01042	.01097	.01261
.234								.00619	.00673	.00729	.00783	.00837	.00947	.01055	.01109	.01273
.250								.00631	.00685	.00741	.00795	.00849	.00959	.01068	.01122	.01286

EXAMPLE: MATERIAL THICKNESS = 1/8", INSIDE RADII = 1/4" R. WHERE THESE CROSS = .00534" IF THE INSIDE OF YOUR BEND IS 20° THE BEND ALLOWANCE (B/A) = .1068 (.00534 X 20°) BEND ALLOWANCE (B/A) PLUS TOTAL OF STRAIGHT LENGTHS = DEVELOPED LENGTH.

337

APPENDIX C-4

BEND ALLOWANCE FOR EACH 1° OF BEND (MILLIMETRE)

Radii / Thickness	0.80	1.58	2.38	3.18	3.96	4.76	5.56	6.35	7.15	7.94	8.74	9.52	11.12	12.70	13.50	15.85
.330	.0161	.0305	.0442	.0580	.0717	.0859	.0996	.1134	.1271	.1413	.1550	.1668	.1967	.2242	.2379	.2796
.406	.0169	.0311	.0448	.0586	.0723	.0865	.1002	.1140	.1277	.1419	.1556	.1694	.1973	.2248	.2385	.2802
.508	.0177	.0319	.0456	.0594	.0731	.0873	.1010	.1148	.1285	.1427	.1564	.1702	.1981	.2256	.2393	.2810
.559	.0181	.0323	.0460	.0598	.0735	.0877	.1014	.1152	.1289	.1431	.1568	.1706	.1985	.2260	.2397	.2814
.635	.0187	.0329	.0466	.0604	.0741	.0883	.1020	.1158	.1295	.1437	.1574	.1712	.1991	.2266	.2403	.2820
.711	.0193	.0335	.0472	.0610	.0747	.0889	.1026	.1164	.1301	.1443	.1580	.1718	.1997	.2272	.2409	.2826
.813	.0201	.0343	.0480	.0617	.0755	.0897	.1034	.1174	.1309	.1451	.1588	.1726	.2005	.2280	.2417	.2834
.965	.0213	.0355	.0492	.0629	.0767	.0909	.1046	.1183	.1321	.1463	.1600	.1737	.2017	.2291	.2429	.2845
1.016	.0217	.0358	.0496	.0633	.0771	.0913	.1050	.1187	.1325	.1467	.1604	.1741	.2021	.2295	.2433	.2849
1.270		.0378	.0516	.0653	.0790	.0932	.1070	.1207	.1345	.1486	.1624	.1761	.2040	.2315	.2453	.2869
1.625		.0406	.0543	.0681	.0818	.0960	.1097	.1235	.1372	.1514	.1652	.1789	.2066	.2343	.2480	.2897
1.829			.0559	.0697	.0834	.0976	.1113	.1251	.1388	.1530	.1667	.1805	.2084	.2359	.2496	.2913
1.981			.0571	.0709	.0846	.0988	.1125	.1263	.1400	.1542	.1679	.1817	.2096	.2371	.2508	.2925
2.058			.0577	.0715	.0852	.0994	.1131	.1269	.1406	.1548	.1685	.1823	.2102	.2377	.2514	.2931
2.311			.0597	.0734	.0872	.1014	.1151	.1288	.1426	.1568	.1705	.1842	.2122	.2396	.2534	.2950
2.388			.0603	.0740	.0878	.1020	.1157	.1294	.1432	.1574	.1711	.1848	.2128	.2402	.2540	.2956
2.591				.0756	.0894	.1035	.1173	.1310	.1448	.1589	.1727	.1864	.2143	.2418	.2556	.2972
2.769				.0770	.0907	.1049	.1187	.1324	.1461	.1603	.1741	.1878	.2157	.2432	.2570	.2986
3.175				.0802	.0939	.1081	.1218	.1356	.1493	.1635	.1772	.1910	.2189	.2464	.2601	.3018
3.962					.1001	.1142	.1280	.1417	.1555	.1696	.1834	.1971	.2250	.2525	.2663	.3079
4.775					.1064	.1206	.1343	.1481	.1618	.1760	.1897	.2035	.2314	.2589	.2726	.3143
5.156						.1235	.1373	.1510	.1648	.1789	.1927	.2064	.2344	.2618	.2756	.3172
5.537								.1540	.1677	.1819	.1957	.2094	.2373	.2648	.2785	.3202
5.944								.1572	.1709	.1851	.1988	.2126	.2405	.2680	.2817	.3234
6.350								.1603	.1741	.1883	.2020	.2157	.2437	.2711	.2849	.3265

EXAMPLE: MATERIAL THICKNESS = 3.175 MM, INSIDE RADII = 6.35 MM R. WHERE THESE CROSS = .1356 MM IF THE INSIDE OF YOUR BEND IS 20° THE BEND ALLOWANCE (B/A) = 2.712 MM (.1356 X 20°) BEND ALLOWANCE (B/A) PLUS TOTAL OF STRAIGHT LENGTHS = DEVELOPED LENGTH

APPENDIX D

GAUGE	THICKNESS		WT. PER SQ. FT.		GAUGE
10	.1406″	3.571 MM	5.625 LBS	2.551 Kg.	10
11	.1250″	3.175 MM	5.000 LBS	2.267 Kg.	11
12	.1094″	2.778 MM	4.375 LBS	1.984 Kg.	12
13	.0938″	2.383 MM	3.750 LBS	1.700 Kg.	13
14	.0781″	1.983 MM	3.125 LBS	1.417 Kg.	14
15	.0703″	1.786 MM	2.813 LBS	1.276 Kg.	15
16	.0625″	1.588 MM	2.510 LBS	1.134 Kg.	16
17	.0563″	1.430 MM	2.250 LBS	1.021 Kg.	17
18	.0500″	1.270 MM	2.000 LBS	0.907 Kg.	18
19	.0438″	1.111 MM	1.750 LBS	0.794 Kg.	19
20	.0375″	0.953 MM	1.500 LBS	0.680 Kg.	20
21	.0344″	0.877 MM	1.375 LBS	0.624 Kg.	21
22	.0313″	0.795 MM	1.250 LBS	0.567 Kg.	22
23	.0280″	0.714 MM	1.125 LBS	0.510 Kg.	23
24	.0250″	0.635 MM	1.000 LBS	0.454 Kg.	24
25	.0219″	0.556 MM	0.875 LBS	0.397 Kg.	25
26	.0188″	0.478 MM	0.750 LBS	0.340 Kg.	26
27	.0172″	0.437 MM	0.687 LBS	0.312 Kg.	27
28	.0156″	0.396 MM	0.625 LBS	0.283 Kg.	28
29	.0141″	0.358 MM	0.563 LBS	0.255 Kg.	29
30	.0120″	0.318 MM	0.500 LBS	0.227 Kg.	30

U.S. STANDARD GAUGES OF SHEET METAL

APPENDIX E

DIMENSION AND SIZE CHART FOR THREADS

Nominal Size		Diameter (Major)		Diameter (Minor)		Tap Drill (For 75% Th'd.)			Threads Per Inch		Pitch (MM)		T.P.I. (Approx.)	
Inch	M.M.	Inch	M.M.	Inch	M.M.	Drill	Inch	M.M.	UNC	UNF	Coarse	Fine	Coarse	Fine
—	M1.4	.055	1.397	—	—	—	—	—	—	—	.3	.2	85	127
0	—	.060	1.524	.0438	1.092	3/64	.0469	1.168	—	80	—	—	—	—
—	M1.6	.063	1.600	—	—	—	—	—	—	—	.35	.2	74	127
1	—	.073	1.854	.0527	1.320	53	.0595	1.499	64	—	—	—	—	—
1	—	.073	1.854	.0550	1.397	53	.0595	1.499	—	72	—	—	—	—
—	M.2	.079	2.006	—	—	—	—	—	—	—	.4	.25	64	101
2	—	.086	2.184	.0628	1.587	50	.0700	1.778	56	—	—	—	—	—
2	—	.086	2.184	.0657	1.651	50	.0700	1.778	—	64	—	—	—	—
—	M2.5	.098	2.489	—	—	—	—	—	—	—	.45	.35	56	74
3	—	.099	2.515	.0719	1.828	47	.0785	1.981	48	—	—	—	—	—
3	—	.099	2.515	.0758	1.905	46	.0810	2.057	—	56	—	—	—	—
4	—	.112	2.845	.0795	2.006	43	.0890	2.261	40	—	—	—	—	—
4	—	.112	2.845	.0849	2.134	42	.0935	2.380	—	48	—	—	—	—
—	M3	.118	2.997	—	—	—	—	—	—	—	.5	.35	51	74
5	—	.125	3.175	.0925	2.336	38	.1015	2.565	40	—	—	—	—	—
5	—	.125	3.175	.9055	2.413	37	.1040	2.641	—	44	—	—	—	—
6	—	.138	3.505	.0975	2.464	36	.1065	2.692	32	—	—	—	—	—
6	—	.138	3.505	.1055	2.667	33	.1130	2.870	—	40	—	—	—	—
—	M4	.157	3.988	—	—	—	—	—	—	—	.7	.35	36	51
8	—	.164	4.166	.1234	3.124	29	.1360	3.454	32	—	—	—	—	—
8	—	.164	4.166	.1279	3.225	29	.1360	3.454	—	36	—	—	—	—
10	—	.190	4.826	.1359	3.429	26	.1470	3.733	24	—	—	—	—	—
10	—	.190	4.826	.1494	3.785	21	.1590	4.038	—	32	—	—	—	—
—	M5	.196	4.978	—	—	—	—	—	—	—	.8	.5	32	51
12	—	.216	5.486	.1619	4.089	16	.1770	4.496	24	—	—	—	—	—
12	—	.216	5.486	.1696	4.293	15	.1800	4.572	—	28	—	—	—	—
—	M6	.236	5.994	—	—	—	—	—	—	—	1.0	.75	25	34
1/4	—	.250	6.350	.1850	4.699	7	.2010	5.105	20	—	—	—	—	—
1/4	—	.250	6.350	.2036	5.156	3	.2130	5.410	—	28	—	—	—	—
5/16	—	.312	7.938	.2403	6.096	F	.2570	6.527	18	—	—	—	—	—
5/16	—	.312	7.938	.2584	6.553	I	.2720	6.908	—	24	—	—	—	—
—	M8	.315	8.001	—	—	—	—	—	—	—	1.25	1.0	20	25
3/8	—	.375	9.525	.2938	7.442	5/16	.3125	7.937	16	—	—	—	—	—
3/8	—	.375	9.525	.3209	8.153	Q	.3320	8.432	—	24	—	—	—	—
—	M10	.393	9.982	—	—	—	—	—	—	—	1.5	1.25	17	20
7/16	—	.437	11.113	.3447	8.738	U	.3680	9.347	14	—	—	—	—	—
7/16	—	.437	11.113	.3726	9.448	25/64	.3906	9.921	—	20	—	—	—	—
—	M12	.471	11.963	—	—	—	—	—	—	—	1.75	1.25	14.5	20
1/2	—	.500	12.700	.4001	10.162	27/64	.4219	10.715	13	—	—	—	—	—
1/2	—	.500	12.700	.4351	11.049	29/64	.4531	11.509	—	20	—	—	—	—
—	M14	.551	13.995	—	—	—	—	—	—	—	2	1.5	12.5	17
9/16	—	.562	14.288	.4542	11.531	31/64	.4844	12.3031	12	—	—	—	—	—
9/16	—	.562	14.288	.4903	12.446	33/64	.5156	13.096	—	18	—	—	—	—
5/8	—	.625	15.875	.5069	12.852	17/32	.5312	13.493	11	—	—	—	—	—
5/8	—	.625	15.875	.5528	14.020	37/64	.5781	14.684	—	18	—	—	—	—
—	M16	.630	16.002	—	—	—	—	—	—	—	2	1.5	12.5	17
—	M18	.709	18.008	—	—	—	—	—	—	—	2.5	1.5	10	17
3/4	—	.750	19.050	.6201	15.748	21/32	.6562	16.668	10	—	—	—	—	—
3/4	—	.750	19.050	.6688	16.967	11/16	.6875	17.462	—	16	—	—	—	—
—	M20	.787	19.990	—	—	—	—	—	—	16	2.5	1.5	10	17
—	M22	.866	21.996	—	—	—	—	—	—	—	2.5	1.5	10	17
7/8	—	.875	22.225	.7307	18.542	49/64	.7656	19.446	9	—	—	—	—	—
7/8	—	.875	22.225	.7822	19.863	13/16	.8125	20.637	—	14	—	—	—	—
—	M24	.945	24.003	—	—	—	—	—	—	—	3	2	8.5	12.5
1	—	1.000	25.400	.8376	21.2598	7/8	.8750	22.225	8	—	—	—	—	—
1	—	1.000	25.400	.8917	22.632	59/64	.9219	23.415	—	12	—	—	—	—
—	M27	1.063	27.000	—	—	—	—	—	—	—	3	2	8.5	12.5

APPENDIX F

VALUES IN THOUSANDTHS OF AN INCH

Nominal Size Range Inches		Class RC1 Precision Sliding			Class RC2 Sliding Fit			Class RC3 Precision Running			Class RC4 Close Running			Class RC5 Medium Running		
		Hole Tol. GR5	Minimum Clearance	Shaft Tol. GR4	Hole Tol. GR6	Minimum Clearance	Shaft Tol. GR5	Hole Tol. GR7	Minimum Clearance	Shaft Tol. GR6	Hole Tol. GR8	Minimum Clearance	Shaft Tol. GR7	Hole Tol. GR8	Minimum Clearance	Shaft Tol. GR7
Over	To	-0		+0	-0		+0	-0		+0	-0		+0	-0		+0
0	.12	+0.15	0.10	-0.12	+0.25	0.10	-0.15	+0.40	0.30	-0.25	+0.60	0.30	-0.40	+0.60	0.60	-0.40
.12	.24	+0.20	0.15	-0.15	+0.30	0.15	-0.20	+0.50	0.40	-0.30	+0.70	0.40	-0.50	+0.70	0.80	-0.50
.24	.40	+0.25	0.20	-0.15	+0.40	0.20	-0.25	+0.60	0.50	-0.40	+0.90	0.50	-0.60	+0.90	1.00	-0.60
.40	.71	+0.30	0.25	-0.20	+0.40	0.25	-0.30	+0.70	0.60	-0.40	+1.00	0.60	-0.70	+1.00	1.20	-0.70
.71	1.19	+0.40	0.30	-0.25	+0.50	0.30	-0.40	+0.80	0.80	-0.50	+1.20	0.80	-0.80	+1.20	1.60	-0.50
1.19	1.97	+0.40	0.40	-0.30	+0.60	0.40	-0.40	+1.00	1.00	-0.60	+1.60	1.00	-1.00	+1.60	2.00	-1.00
1.97	3.15	+0.50	0.40	-0.30	+0.70	0.40	-0.50	+1.20	1.20	-0.70	+1.80	1.20	-1.20	+1.80	2.50	-1.20
3.15	4.73	+0.60	0.50	-0.40	+0.90	0.50	-0.60	+1.40	1.40	-0.90	+2.20	1.40	-1.40	+2.20	3.00	-1.40
4.73	7.09	+0.70	0.60	-0.50	+1.00	0.60	-0.70	+1.60	1.60	-1.00	+2.50	1.60	-1.60	+2.50	3.50	-1.60
7.09	9.85	+0.80	0.60	-0.60	+1.20	0.60	-0.80	+1.80	2.00	-1.20	+2.80	2.00	-1.80	+2.80	4.50	-1.80
9.85	12.41	+0.90	0.80	-0.60	+1.20	1.00	-0.90	+2.00	2.50	-1.20	+3.00	2.50	-2.00	+3.00	5.00	-2.00
12.41	15.75	+1.00	1.00	-0.70	+1.40	1.00	-1.00	+2.20	3.00	-1.40	+3.50	3.00	-2.20	+3.50	6.00	-2.20

Nominal Size Range Inches		Class RC6 Medium Running			Class RC7 Free Running			Class RC8 Loose Running			Class RC9 Loose Running		
		Hole Tol. GR9	Minimum Clearance	Shaft Tol. GR8	Hole Tol. GR9	Minimum Clearance	Shaft Tol. GR8	Hole Tol. GR10	Minimum Clearance	Shaft Tol. GR9	Hole Tol. GR11	Minimum Clearance	Shaft Tol. GR10
Over	To	-0		+0	-0		+0	-0		+0	-0		+0
0	.12	+1.00	0.60	-0.60	+1.00	1.00	-0.60	+1.60	2.50	-1.00	+2.50	4.00	-1.60
.12	.24	+1.20	0.80	-0.70	+1.20	1.20	-0.70	+1.80	2.80	-1.20	+3.00	4.50	-1.80
.24	.40	+1.40	1.00	-0.90	+1.40	1.60	-0.90	+2.20	3.00	-1.40	+3.50	6.00	-2.20
.40	.71	+1.60	1.20	-1.00	+1.60	2.00	-1.00	+2.80	3.50	-1.60	+4.00	6.00	-2.80
.71	1.19	+2.00	1.60	-1.20	+2.00	2.50	-1.20	+3.50	4.50	-2.00	+5.00	7.00	-3.50
1.19	1.97	+2.50	2.00	-1.60	+2.50	3.00	-1.60	+4.00	5.00	-2.50	+6.00	8.00	-4.00
1.97	3.15	+3.00	2.50	-1.80	+3.00	4.00	-1.80	+4.50	6.00	-3.00	+7.00	9.00	-4.50
3.15	4.73	+3.50	3.00	-2.20	+3.50	5.00	-2.20	+5.00	7.00	-3.50	+9.00	10.00	-5.00
4.73	7.09	+4.00	3.50	-2.50	+4.00	6.00	-2.50	+6.00	8.00	-4.00	+10.00	12.00	-6.00
7.09	9.85	+4.50	4.00	-2.80	+4.50	7.00	-2.80	+7.00	10.00	-4.50	+12.00	15.00	-7.00
9.85	12.41	+5.00	5.00	-3.00	+5.00	8.00	-3.00	+8.00	12.00	-5.00	+12.00	18.00	-8.00
12.41	15.75	+6.00	6.00	-3.50	+6.00	10.00	-3.50	+9.00	14.00	-6.00	+14.00	22.00	-9.00

VALUES IN MILLIMETRES

Nominal Size Range Millimetres		Class RC1 Precision Sliding			Class RC2 Sliding Fit			Class RC3 Precision Running			Class RC4 Close Running			Class RC5 Medium Running		
		Hole Tol. H5	Minimum Clearance	Shaft Tol. g4	Hole Tol. H6	Minimum Clearance	Shaft Tol. g5	Hole Tol. H7	Minimum Clearance	Shaft Tol. f6	Hole Tol. H8	Minimum Clearance	Shaft Tol. f7	Hole Tol. H8	Minimum Clearance	Shaft Tol. e7
Over	To	-0		+0	-0		+0	-0		+0	-0		+0	-0		+0
0	3	+0.004	0.003	-0.003	+0.006	0.003	-0.004	+0.010	0.008	-0.006	+0.015	0.008	-0.010	+0.015	0.015	-0.010
3	6	+0.005	0.004	-0.004	+0.008	0.004	-0.005	+0.013	0.010	-0.008	+0.018	0.010	-0.013	+0.018	0.020	-0.013
6	10	+0.006	0.005	-0.004	+0.010	0.005	-0.006	+0.015	0.013	-0.010	+0.023	0.013	-0.015	+0.023	0.025	-0.015
10	18	+0.008	0.006	-0.005	+0.010	0.006	-0.008	+0.018	0.015	-0.010	+0.025	0.015	-0.018	+0.025	0.030	-0.018
18	30	+0.010	0.008	-0.006	+0.013	0.008	-0.010	+0.020	0.020	-0.013	+0.030	0.020	-0.020	+0.030	0.040	-0.020
30	50	+0.010	0.010	-0.008	+0.015	0.010	-0.010	+0.030	0.030	-0.015	+0.040	0.030	-0.030	+0.040	0.050	-0.030
50	80	+0.013	0.010	-0.008	+0.018	0.010	-0.013	+0.030	0.030	-0.020	+0.050	0.030	-0.030	+0.050	0.060	-0.030
80	120	+0.015	0.013	-0.010	+0.023	0.013	-0.015	+0.040	0.040	-0.020	+0.060	0.040	-0.040	+0.060	0.080	-0.040
120	180	+0.018	0.015	-0.013	+0.025	0.015	-0.018	+0.040	0.040	-0.030	+0.060	0.040	-0.040	+0.060	0.090	-0.040
180	250	+0.020	0.015	-0.015	+0.030	0.015	-0.020	+0.050	0.050	-0.030	+0.070	0.050	-0.050	+0.070	0.110	-0.050
250	315	+0.023	0.020	-0.015	+0.030	0.020	-0.023	+0.050	0.060	-0.030	+0.080	0.060	-0.050	+0.080	0.130	-0.050
315	400	+0.025	0.025	-0.018	+0.036	0.025	-0.025	+0.060	0.080	-0.040	+0.090	0.080	-0.060	+0.090	0.150	-0.060

Nominal Size Range Millimetres		Class RC6 Medium Running			Class RC7 Free Running			Class RC8 Loose Running			Class RC9 Loose Running		
		Hole Tol. H9	Minimum Clearance	Shaft Tol. e8	Hole Tol. H9	Minimum Clearance	Shaft Tol. d8	Hole Tol. H10	Minimum Clearance	Shaft Tol. e9	Hole Tol. GR11	Minimum Clearance	Shaft Tol. gr10
Over	To	-0		+0	-0		+0	-0		+0	-0		+0
0	3	+0.025	0.015	-0.015	+0.025	0.025	-0.015	+0.041	0.064	-0.025	+0.060	0.100	-0.040
3	6	+0.030	0.015	-0.018	+0.030	0.030	-0.018	+0.046	0.071	-0.030	+0.080	0.110	-0.050
6	10	+0.036	0.025	-0.023	+0.036	0.040	-0.023	+0.056	0.076	-0.036	+0.070	0.130	-0.060
10	18	+0.040	0.030	-0.025	+0.040	0.050	-0.025	+0.070	0.090	-0.040	+0.100	0.150	-0.070
18	30	+0.050	0.040	-0.030	+0.050	0.060	-0.030	+0.090	0.110	-0.050	+0.130	0.180	-0.090
30	50	+0.060	0.050	-0.040	+0.060	0.080	-0.040	+0.100	0.130	-0.060	+0.150	0.200	-0.100
50	80	+0.080	0.060	-0.050	+0.080	0.100	-0.050	+0.110	0.150	-0.080	+0.180	0.230	-0.120
80	120	+0.090	0.080	-0.060	+0.090	0.130	-0.060	+0.130	0.180	-0.090	+0.230	0.250	-0.130
120	180	+0.100	0.090	-0.060	+0.100	0.150	-0.060	+0.150	0.200	-0.100	+0.250	0.300	-0.150
180	250	+0.110	0.100	-0.070	+0.110	0.180	-0.070	+0.180	0.250	-0.110	+0.300	0.380	-0.180
250	315	+0.130	0.130	-0.080	+0.130	0.200	-0.080	+0.200	0.300	-0.130	+0.300	0.460	-0.200
315	400	+0.150	0.150	-0.090	+0.150	0.250	-0.090	+0.230	0.360	-0.150	+0.360	0.560	-0.230

Running and sliding fits

VALUES IN THOUSANDTHS OF AN INCH

Nominal Size Range Inches		Class LC1			Class LC2			Class LC3			Class LC4			Class LC5			Class LC6		
		Hole Tol. GR6	Minimum Clearance	Shaft Tol. GR5	Hole Tol. GR8	Minimum Clearance	Shaft Tol. GR7	Hole Tol. GR10	Minimum Clearance	Shaft Tol. GR9	Hole Tol. GR7	Minimum Clearance	Shaft Tol. GR6	Hole Tol. GR9	Minimum Clearance	Shaft Tol. GR8	Hole Tol. GR9	Minimum Clearance	Shaft Tol. GR8
Over	To	-0		+0	-0		+0	-0		+0	-0		+0	-0		+0	-0		+0
0	.12	+0.25	0	-0.15	+0.4	0	-0.25	+0.6	0	-0.4	+1.6	0	-1.0	+0.4	0.10	-0.25	+1.0	0.3	-0.6
.12	.24	+0.30	0	-0.20	+0.5	0	-0.30	+0.7	0	-0.5	+1.8	0	-1.2	+0.5	0.15	-0.30	+1.2	0.4	-0.7
.24	.40	+0.40	0	-0.25	+0.6	0	-0.40	+0.9	0	-0.6	+2.2	0	-1.4	+0.6	0.20	-0.40	+1.4	0.5	-0.9
.40	.71	+0.40	0	-0.30	+0.7	0	-0.40	+1.0	0	-0.7	+2.8	0	-1.6	+0.7	0.25	-0.40	+1.6	0.6	-1.0
.71	1.19	+0.50	0	-0.40	+0.8	0	-0.50	+1.2	0	-0.8	+3.5	0	-2.0	+0.8	0.30	-0.50	+2.0	0.8	-1.2
1.19	1.97	+0.60	0	-0.40	+1.0	0	-0.60	+1.6	0	-1.0	+4.0	0	-2.5	+1.0	0.40	-0.60	+2.5	1.0	-1.6
1.97	3.15	+0.70	0	-0.50	+1.2	0	-0.70	+1.8	0	-1.2	+4.5	0	-3.0	+1.2	0.40	-0.70	+3.0	1.2	-1.8
3.15	4.73	+0.90	0	-0.60	+1.4	0	-0.90	+2.7	0	-1.4	+5.0	0	-3.5	+1.4	0.50	-0.90	+3.5	1.4	-2.2
4.73	7.09	+1.00	0	-0.70	+1.6	0	-1.00	+2.5	0	-1.6	+6.0	0	-4.0	+1.6	0.60	-1.00	+4.0	1.6	-2.5
7.09	9.85	+1.20	0	-0.80	+1.8	0	-1.20	+2.8	0	-1.8	+7.0	0	-4.5	+1.8	0.60	-1.20	+4.5	2.0	-2.8
9.85	12.41	+1.20	0	-0.90	+2.0	0	-1.20	+3.0	0	-2.0	+8.0	0	-5.0	+2.0	0.70	-1.20	+5.0	2.2	-3.0
12.41	15.75	+1.40	0	-1.00	+2.2	0	-1.40	+3.5	0	-2.2	+9.0	0	-6.0	+2.2	0.70	-1.40	+6.0	2.5	-3.5

Nominal Size Range Inches		Class LC7			Class LC8			Class LC9			Class LC10			Class LC11		
		Hole Tol. GR10	Minimum Clearance	Shaft Tol. GR9	Hole Tol. GR10	Minimum Clearance	Shaft Tol. GR9	Hole Tol. GR11	Minimum Clearance	Shaft Tol. GR10	Hole Tol. GR12	Minimum Clearance	Shaft Tol. GR11	Hole Tol. GR13	Minimum Clearance	Shaft Tol. GR12
Over	To	-0		+0	-0		+0	-0		+0	-0		+0	-0		+0
0	.12	+1.6	0.6	-1.0	+1.6	1.0	-1.0	+2.5	2.5	-1.6	+4.0	4.0	-2.5	+6.0	5.0	-4.0
.12	.24	+1.8	0.8	-1.2	+1.8	1.2	-1.2	+3.0	2.8	-1.8	+5.0	4.5	-3.0	+7.0	6.0	-5.0
.24	.40	+2.2	1.0	-1.4	+2.2	1.6	-1.4	+3.5	3.0	-2.2	+6.0	5.0	-3.5	+9.0	7.0	-6.0
.40	.71	+2.8	1.2	-1.6	+2.8	2.0	-1.6	+5.0	3.5	-2.8	+7.0	6.0	-4.0	+10.0	8.0	-7.0
.71	1.19	+3.5	1.6	-2.0	+3.5	2.5	-2.0	+5.0	4.5	-3.5	+8.0	7.0	-5.0	+12.0	10.0	-7.0
1.19	1.97	+4.0	2.0	-2.5	+4.0	3.6	-2.5	+6.0	5.0	-4.0	+10.0	8.0	-6.0	+16.0	12.0	-10.0
1.97	3.15	+4.5	2.5	-3.0	+4.5	4.0	-3.0	+7.0	6.0	-4.5	+12.0	10.0	-7.0	+18.0	14.0	-12.0
3.15	4.73	+5.0	3.0	-3.5	+5.0	5.0	-3.5	+9.0	7.0	-5.0	+14.0	11.0	-9.0	+22.0	16.0	-14.0
4.73	7.09	+6.0	3.5	-4.0	+6.0	6.0	-4.0	+10.0	8.0	-6.0	+16.0	12.0	-10.0	+25.0	18.0	-16.0
7.09	9.85	+7.0	4.0	-4.5	+7.0	7.0	-4.5	+12.0	10.0	-7.0	+18.0	16.0	-12.0	+28.0	22.0	-18.0
9.85	12.41	+8.0	4.5	-5.0	+8.0	7.0	-5.0	+12.0	12.0	-8.0	+20.0	20.0	-12.0	+30.0	28.0	-20.0
12.41	15.75	+9.0	5.0	-6.0	+9.0	8.0	-6.0	+14.0	14.0	-9.0	+22.0	22.0	-14.0	+35.0	30.0	-22.0

VALUES IN MILLIMETRES

Nominal Size Range Millimetres		Class LC1			Class LC2			Class LC3			Class LC4			Class LC5			Class LC6		
		Hole Tol. H6	Minimum Clearance	Shaft Tol. h5	Hole Tol. H7	Minimum Clearance	Shaft Tol. h6	Hole Tol. H8	Minimum Clearance	Shaft Tol. h7	Hole Tol. H10	Minimum Clearance	Shaft Tol. h9	Hole Tol. H7	Minimum Clearance	Shaft Tol. g6	Hole Tol. H9	Minimum Clearance	Shaft Tol. f8
Over	To	-0		+0	-0		+0	-0		+0	-0		+0	-0		+0	-0		+0
0	3	+0.006	0	-0.004	+0.010	0	-0.006	+0.015	0	-0.010	+0.041	0	-0.025	+0.010	0.002	-0.006	+0.025	0.008	-0.015
3	6	+0.008	0	-0.005	+0.013	0	-0.008	+0.018	0	-0.013	+0.046	0	-0.030	+0.013	0.004	-0.008	+0.030	0.010	-0.018
6	10	+0.010	0	-0.006	+0.015	0	-0.010	+0.023	0	-0.015	+0.056	0	-0.036	+0.015	0.005	-0.010	+0.036	0.013	-0.023
10	18	+0.010	0	-0.008	+0.018	0	-0.010	+0.025	0	-0.018	+0.070	0	-0.040	+0.018	0.006	-0.010	+0.041	0.015	-0.025
18	30	+0.013	0	-0.010	+0.020	0	-0.013	+0.030	0	-0.020	+0.090	0	-0.050	+0.020	0.008	-0.013	+0.050	0.020	-0.030
30	50	+0.015	0	-0.010	+0.025	0	-0.015	+0.041	0	-0.025	+0.100	0	-0.060	+0.025	0.010	-0.015	+0.060	0.030	-0.040
50	80	+0.018	0	-0.013	+0.030	0	-0.018	+0.046	0	-0.030	+0.110	0	-0.080	+0.030	0.010	-0.018	+0.080	0.030	-0.050
80	120	+0.023	0	-0.015	+0.036	0	-0.023	+0.056	0	-0.036	+0.130	0	-0.080	+0.036	0.013	-0.023	+0.090	0.040	-0.060
120	180	+0.025	0	-0.018	+0.041	0	-0.025	+0.064	0	-0.041	+0.150	0	-0.100	+0.041	0.015	-0.025	+0.100	0.040	-0.060
180	250	+0.030	0	-0.020	+0.046	0	-0.030	+0.071	0	-0.046	+0.180	0	-0.110	+0.046	0.015	-0.030	+0.110	0.050	-0.070
250	315	+0.020	0	-0.023	+0.051	0	-0.030	+0.076	0	-0.051	+0.200	0	-0.130	+0.051	0.018	-0.030	+0.130	0.060	-0.080
315	400	+0.036	0	-0.025	+0.056	0	-0.036	+0.089	0	-0.056	+0.230	0	-0.150	+0.056	0.018	-0.036	+0.150	0.060	-0.090

Nominal Size Range Millimetres		Class LC7			Class LC8			Class LC9			Class LC10			Class LC11		
		Hole Tol. H10	Minimum Clearance	Shaft Tol. e9	Hole Tol. H10	Minimum Clearance	Shaft Tol. d9	Hole Tol. H11	Minimum Clearance	Shaft Tol. c10	Hole Tol. GR12	Minimum Clearance	Shaft Tol. gr11	Hole Tol. GR13	Minimum Clearance	Shaft Tol. gr12
Over	To	-0		+0	-0		+0	-0		+0	-0		+0	-0		+0
0	3	+0.041	0.015	-0.025	+0.041	0.025	-0.025	+0.064	0.06	-0.041	+0.10	0.10	-0.06	+0.15	0.13	-0.10
3	6	+0.046	0.020	-0.030	+0.046	0.030	-0.030	+0.076	0.07	-0.046	+0.13	0.11	-0.08	+0.18	0.15	-0.13
6	10	+0.056	0.025	-0.036	+0.056	0.041	-0.036	+0.089	0.08	-0.056	+0.15	0.13	-0.09	+0.23	0.18	-0.15
10	18	+0.070	0.030	-0.040	+0.070	0.050	-0.040	+0.100	0.09	-0.070	+0.18	0.15	-0.10	+0.25	0.20	-0.18
18	30	+0.090	0.040	-0.050	+0.090	0.060	-0.050	+0.130	0.11	-0.090	+0.20	0.18	-0.13	+0.31	0.25	-0.20
30	50	+0.100	0.050	-0.060	+0.100	0.090	-0.060	+0.150	0.13	-0.100	+0.25	0.20	-0.15	+0.41	0.31	-0.25
50	80	+0.110	0.060	-0.080	+0.110	0.100	-0.080	+0.180	0.15	-0.110	+0.31	0.25	-0.18	+0.46	0.36	-0.31
80	120	+0.130	0.080	-0.090	+0.130	0.130	-0.090	+0.230	0.18	-0.130	+0.36	0.28	-0.23	+0.56	0.41	-0.36
120	180	+0.150	0.090	-0.100	+0.150	0.150	-0.100	+0.250	0.20	-0.150	+0.41	0.31	-0.25	+0.64	0.46	-0.41
180	250	+0.180	0.100	-0.110	+0.180	0.180	-0.110	+0.310	0.25	-0.180	+0.46	0.41	-0.31	+0.71	0.56	-0.46
250	315	+0.200	0.110	-0.130	+0.200	0.180	-0.130	+0.310	0.31	-0.200	+0.51	0.51	-0.31	+0.76	0.71	-0.51
315	400	+0.230	0.130	-0.150	+0.230	0.200	-0.150	+0.360	0.36	-0.230	+0.56	0.56	-0.36	+0.89	0.76	-0.56

Locational clearance fits

VALUES IN THOUSANDTHS OF AN INCH

Nominal Size Range Inches		Class LT1			Class LT2			Class LT3			Class LT4			Class LT5			Class LT6		
		Hole Tol. GR7	Maximum Interference	Shaft Tol. GR6	Hole Tol. GR8	Maximum Interference	Shaft Tol. GR7	Hole Tol. GR7	Maximum Interference	Shaft Tol. GR7	Hole Tol. GR8	Maximum Interference	Shaft Tol. GR6	Hole Tol. GR7	Maximum Interference	Shaft Tol. GR7	Hole Tol. GR8	Maximum Interference	Shaft Tol. GR7
Over	To	-0		+0	-0		+0	-0		+0	-0		+0	-0		+0	-0		+0
0	.12	+0.4	0.10	-0.25	+0.6	0.20	-0.4	+0.4	0.25	-0.25	+0.6	0.4	-0.4	+0.4	0.5	-0.25	+0.6	0.65	-0.4
.12	.24	+0.5	0.15	-0.30	+0.7	0.25	-0.5	+0.5	0.40	-0.30	+0.7	0.6	-0.5	+0.5	0.6	-0.30	+0.7	0.80	-0.5
.24	.40	+0.6	0.20	-0.40	+0.9	0.30	-0.6	+0.6	0.50	-0.40	+0.9	0.7	-0.6	+0.6	0.8	-0.40	+0.9	1.00	-0.6
.40	.71	+0.7	0.20	-0.40	+1.0	0.30	-0.7	+0.7	0.50	-0.40	+1.0	0.8	-0.7	+0.7	0.9	-0.40	+1.0	1.20	-0.7
.71	1.19	+0.8	0.25	-0.50	+1.2	0.40	-0.8	+0.8	0.60	-0.50	+1.2	0.9	-0.8	+0.8	1.1	-0.50	+1.2	1.40	-0.8
1.19	1.97	+1.0	0.30	-0.60	+1.6	0.50	-1.0	+1.0	0.70	-0.60	+1.6	1.1	-1.0	+1.0	1.3	-0.60	+1.6	1.70	-1.0
1.97	3.15	+1.2	0.30	-0.70	+1.8	0.60	-1.2	+1.2	0.80	-0.70	+1.8	1.3	-1.2	+1.2	1.5	-0.70	+1.8	2.00	-1.2
3.15	4.73	+1.4	0.40	-0.90	+2.2	0.70	-1.4	+1.4	1.00	-0.90	+2.2	1.5	-1.4	+1.4	1.9	-0.90	+2.2	2.40	-1.4
4.73	7.09	+1.6	0.50	-1.00	+2.5	0.80	-1.6	+1.6	1.10	-1.00	+2.5	1.7	-1.6	+1.6	2.2	-1.00	+2.5	2.80	-1.6
7.09	9.85	+1.8	0.60	-1.20	+2.8	0.90	-1.8	+1.8	1.40	-1.20	+2.8	2.0	-1.8	+1.8	2.6	-1.20	+2.8	3.20	-1.8
9.85	12.41	+2.0	0.60	-1.20	+3.0	1.00	-2.0	+2.0	1.40	-1.20	+3.0	2.2	-2.0	+2.0	2.6	-1.20	+3.0	3.40	-2.0
12.41	15.75	+2.2	0.70	-1.40	+3.5	1.00	-2.2	+2.2	1.60	-1.40	+3.5	2.4	-2.2	+2.2	3.0	-1.40	+3.5	3.80	-2.2

VALUES IN MILLIMETRES

Nominal Size Range Millimetres		Class LT1			Class LT2			Class LT3			Class LT4			Class LT5			Class LT6		
		Hole Tol. H7	Maximum Interference	Shaft Tol. js6	Hole Tol. H8	Maximum Interference	Shaft Tol. js7	Hole Tol. H7	Maximum Interference	Shaft Tol. k6	Hole Tol. H8	Maximum Interference	Shaft Tol. k7	Hole Tol. H7	Maximum Interference	Shaft Tol. n6	Hole Tol. H8	Maximum Interference	Shaft Tol. n7
Over	To	-0		+0	-0		+0	-0		+0	-0		+0	-0		+0	-0		+0
0	3	+0.010	0.002	-0.006	+0.015	0.005	-0.010	+0.010	0.006	-0.006	+0.015	0.010	-0.010	+0.010	0.013	-0.006	+0.015	0.016	-0.010
3	6	+0.013	0.004	-0.008	+0.018	0.006	-0.013	+0.013	0.010	-0.008	+0.018	0.015	-0.013	+0.013	0.015	-0.008	+0.018	0.020	-0.013
6	10	+0.015	0.005	-0.010	+0.023	0.008	-0.015	+0.015	0.013	-0.010	+0.023	0.018	-0.015	+0.015	0.020	-0.010	+0.023	0.025	-0.018
10	18	+0.018	0.005	-0.010	+0.025	0.008	-0.018	+0.018	0.013	-0.010	+0.025	0.020	-0.018	+0.020	0.028	-0.013	+0.030	0.036	-0.020
18	30	+0.020	0.006	-0.013	+0.030	0.010	-0.020	+0.020	0.015	-0.013	+0.041	0.028	-0.025	+0.025	0.033	-0.015	+0.041	0.044	-0.025
30	50	+0.025	0.008	-0.015	+0.041	0.013	-0.025	+0.025	0.020	-0.015	+0.046	0.033	-0.030	+0.030	0.038	-0.018	+0.046	0.051	-0.030
50	80	+0.030	0.008	-0.018	+0.046	0.015	-0.030	+0.030	0.020	-0.018	+0.056	0.038	-0.036	+0.036	0.048	-0.023	+0.056	0.062	-0.036
80	120	+0.036	0.010	-0.023	+0.056	0.018	-0.036	+0.036	0.025	-0.023	+0.064	0.044	-0.041	+0.041	0.056	-0.025	+0.064	0.071	-0.041
120	180	+0.041	0.013	-0.025	+0.064	0.020	-0.041	+0.041	0.028	-0.025	+0.071	0.051	-0.046	+0.046	0.066	-0.030	+0.071	0.081	-0.046
180	250	+0.046	0.015	-0.030	+0.071	0.023	-0.046	+0.046	0.036	-0.030	+0.076	0.056	-0.051	+0.051	0.066	-0.030	+0.076	0.086	-0.051
250	315	+0.051	0.015	-0.030	+0.076	0.025	-0.051	+0.051	0.036	-0.030	+0.089	0.062	-0.056	+0.056	0.076	-0.036	+0.089	0.096	-0.056
315	400	+0.056	0.018	-0.036	+0.089	0.025	-0.056	+0.056	0.041	-0.036									

Location transition fits

VALUES IN THOUSANDTHS OF AN INCH

Nominal Size Range Inches		Class LN1 Light Press Fit			Class LN2 Medium Press Fit			Class LN3 Heavy Press Fit			Class LN4			Class LN5			Class LN6		
		Hole Tol. GR6	Maximum Interference	Shaft Tol. GR5	Hole Tol. GR7	Maximum Interference	Shaft Tol. GR6	Hole Tol. GR7	Maximum Interference	Shaft Tol. GR6	Hole Tol. GR8	Maximum Interference	Shaft Tol. GR7	Hole Tol. GR9	Maximum Interference	Shaft Tol. GR8	Hole Tol. GR10	Maximum Interference	Shaft Tol. GR9
Over	To	-0		+0	-0		+0	-0		+0	-0		+0	-0		+0	-0		+0
0	.12	+0.25	0.40	-0.15	+0.4	0.65	-0.25	+0.4	0.75	-0.25	+0.6	1.2	-0.4	+1.0	1.8	-0.6	+1.6	3.0	-1.0
.12	.24	+0.30	0.50	-0.20	+0.5	0.80	-0.30	+0.5	0.90	-0.30	+0.7	1.5	-0.5	+1.2	2.3	-0.7	+1.8	3.6	-1.2
.24	.40	+0.40	0.65	-0.25	+0.6	1.00	-0.40	+0.6	1.20	-0.40	+0.9	1.8	-0.6	+1.4	2.8	-0.9	+2.2	4.4	-1.4
.40	.71	+0.40	0.70	-0.30	+0.7	1.10	-0.40	+0.7	1.40	-0.40	+1.0	2.2	-0.7	+1.6	3.4	-1.0	+2.8	5.6	-1.6
.71	1.19	+0.50	0.90	-0.40	+0.8	1.30	-0.50	+0.8	1.70	-0.50	+1.2	2.6	-0.8	+2.0	4.2	-1.2	+3.5	7.0	-2.0
1.19	1.97	+0.60	1.00	-0.40	+1.0	1.60	-0.60	+1.0	2.00	-0.60	+1.6	3.4	-1.0	+2.5	5.3	-1.6	+4.0	8.5	-2.5
1.97	3.15	+0.70	1.30	-0.50	+1.2	2.10	-0.70	+1.2	2.30	-0.70	+2.2	4.8	-1.4	+4.0	7.7	-2.2	+5.0	11.5	-3.5
3.15	4.73	+0.90	1.60	-0.60	+1.4	2.50	-0.90	+1.4	2.90	-0.90	+2.5	5.6	-1.6	+4.5	8.7	-2.5	+6.0	13.5	-4.0
4.73	7.09	+1.00	1.90	-0.70	+1.6	2.80	-1.00	+1.6	3.50	-1.00	+2.8	6.6	-1.8	+5.0	10.3	-2.8	+7.0	16.5	-4.5
7.09	9.85	+1.20	2.20	-0.80	+1.8	3.20	-1.20	+1.8	4.20	-1.20	+3.0	7.5	-2.0	+6.0	12.0	-3.0	+8.0	19.0	-5.0
9.85	12.41	+1.20	2.30	-0.90	+2.0	3.40	-1.20	+2.0	4.70	-1.20	+3.0	7.5	-2.0	+6.0	12.0	-3.0	+8.0	19.0	-5.0
12.41	15.75	+1.40	2.60	-1.00	+2.2	3.90	-1.40	+2.2	5.90	-1.40	+3.5	8.7	-2.2	+6.0	14.5	-3.5	+9.0	23.0	-6.0

VALUES IN MILLIMETRES

Nominal Size Range Millimetres		Class LN1 Light Press Fit			Class LN2 Medium Press Fit			Class LN3 Heavy Press Fit			Class LN4			Class LN5			Class LN6		
		Hole Tol. GR6	Maximum Interference	Shaft Tol. gr5	Hole Tol. H7	Maximum Interference	Shaft Tol. p6	Hole Tol. H7	Maximum Interference	Shaft Tol. t6	Hole Tol. GR8	Maximum Interference	Shaft Tol. gr7	Hole Tol. GR9	Maximum Interference	Shaft Tol. gr8	Hole Tol. GR10	Maximum Interference	Shaft Tol. gr9
Over	To	-0		+0	-0		+0	-0		+0	-0		+0	-0		+0	-0		+0
0	3	+0.006		-0.004	+0.010	0.016	-0.006	+0.010	0.019	-0.006	+0.015	0.030	-0.010	+0.025	0.046	-0.015	+0.041	0.076	-0.025
3	6	+0.008		-0.005	+0.013	0.020	-0.008	+0.013	0.023	-0.008	+0.018	0.038	-0.013	+0.030	0.059	-0.018	+0.046	0.091	-0.030
6	10	+0.010		-0.006	+0.015	0.025	-0.010	+0.015	0.030	-0.010	+0.023	0.046	-0.015	+0.036	0.071	-0.023	+0.056	0.112	-0.036
10	18	+0.010		-0.008	+0.018	0.028	-0.010	+0.018	0.036	-0.010	+0.025	0.056	-0.018	+0.041	0.086	-0.025	+0.071	0.142	-0.041
18	30	+0.013		-0.010	+0.020	0.033	-0.013	+0.020	0.044	-0.013	+0.030	0.066	-0.020	+0.051	0.107	-0.030	+0.089	0.178	-0.051
30	50	+0.015		-0.010	+0.025	0.041	-0.015	+0.025	0.051	-0.015	+0.041	0.086	-0.025	+0.064	0.135	-0.041	+0.102	0.216	-0.064
50	80	+0.018		-0.013	+0.030	0.054	-0.018	+0.030	0.059	-0.018	+0.046	0.102	-0.030	+0.076	0.160	-0.046	+0.114	0.254	-0.076
80	120	+0.023		-0.015	+0.036	0.064	-0.023	+0.036	0.074	-0.023	+0.056	0.122	-0.036	+0.102	0.196	-0.056	+0.127	0.292	-0.102
120	180	+0.025		-0.018	+0.041	0.071	-0.025	+0.041	0.089	-0.025	+0.064	0.142	-0.041	+0.114	0.221	-0.064	+0.152	0.343	-0.114
180	250	+0.030		-0.020	+0.046	0.081	-0.030	+0.046	0.107	-0.030	+0.071	0.168	-0.046	+0.127	0.262	-0.071	+0.178	0.419	-0.127
250	315	+0.030		-0.023	+0.051	0.086	-0.030	+0.051	0.119	-0.030	+0.076	0.191	-0.051	+0.152	0.305	-0.076	+0.203	0.483	-0.152
315	400	+0.036		-0.025	+0.056	0.099	-0.036	+0.056	0.150	-0.036	+0.089	0.221	-0.056	+0.152	0.368	-0.089	+0.229	0.584	-0.152

Locational interference fits

VALUES IN THOUSANDTHS OF AN INCH

Nominal Size Range Inches		Class FN1 Light Drive Fit			Class FN2 Medium Drive Fit			Class FN3 Heavy Drive Fit			Class FN4 Shrink Fit			Class FN5 Heavy Shrink Fit		
		Hole Tol. GR6	Maximum Interference	Shaft Tol. GR5	Hole Tol. GR7	Maximum Interference	Shaft Tol. GR6	Hole Tol. GR6	Maximum Interference	Shaft Tol. GR6	Hole Tol. GR7	Maximum Interference	Shaft Tol. GR6	Hole Tol. GR8	Maximum Interference	Shaft Tol. GR7
Over	To	-0		+0	-0		+0	-0		+0	-0		+0	-0		+0
0	.12	+0.25	0.50	-0.15	+0.40	0.85	-0.25				+0.40	0.95	-0.25	+0.60	1.30	-0.40
.12	.24	+0.30	0.60	-0.20	+0.50	1.00	-0.30				+0.50	1.20	-0.30	+0.70	1.70	-0.50
.24	.40	+0.40	0.75	-0.25	+0.60	1.40	-0.40				+0.60	1.60	-0.40	+0.90	2.00	-0.60
.40	.56	+0.40	0.80	-0.30	+0.70	1.60	-0.40				+0.70	1.80	-0.40	+1.00	2.30	-0.70
.56	.71	+0.40	0.90	-0.30	+0.70	1.60	-0.40				+0.70	1.80	-0.40	+1.00	2.50	-0.70
.71	.95	+0.50	1.10	-0.40	+0.80	1.90	-0.50				+0.80	2.10	-0.50	+1.20	3.00	-0.80
.95	1.19	+0.50	1.20	-0.40	+0.80	1.90	-0.50	+0.80	2.10	-0.50	+0.80	2.30	-0.50	+1.20	3.30	-0.80
1.19	1.58	+0.60	1.30	-0.40	+1.00	2.40	-0.60	+1.00	2.60	-0.60	+1.00	3.10	-0.60	+1.60	4.00	-1.00
1.58	1.97	+0.60	1.40	-0.40	+1.00	2.40	-0.60	+1.00	2.80	-0.60	+1.00	3.40	-0.60	+1.60	5.00	-1.00
1.97	2.56	+0.70	1.80	-0.50	+1.20	2.70	-0.70	+1.20	3.20	-0.70	+1.20	4.20	-0.70	+1.80	6.20	-1.20
2.56	3.15	+0.70	1.90	-0.50	+1.20	2.90	-0.70	+1.20	3.70	-0.70	+1.20	4.70	-0.70	+1.80	7.20	-1.20
3.15	3.94	+0.90	2.40	-0.60	+1.40	3.70	-0.90	+1.40	4.40	-0.70	+1.40	5.90	-0.90	+2.20	8.40	-1.40

VALUES IN MILLIMETRES

Nominal Size Range Millimetres		Class FN1 Light Drive Fit			Class FN2 Medium Drive Fit			Class FN3 Heavy Drive Fit			Class FN4 Shrink Fit			Class FN5 Heavy Shrink Fit		
		Hole Tol. GR6	Maximum Interference	Shaft Tol. gr5	Hole Tol. H7	Maximum Interference	Shaft Tol. s6	Hole Tol. H7	Maximum Interference	Shaft Tol. t6	Hole Tol. GR8	Maximum Interference	Shaft Tol. GR6	Hole Tol. H8	Maximum Interference	Shaft Tol. t7
Over	To	-0		+0	-0		+0	-0		+0	-0		+0	-0		+0
0	3	+0.006	0.013	-0.004	+0.010	0.216	-0.006				+0.010	0.024	-0.006	+0.015	0.033	-0.010
3	6	+0.007	0.015	-0.005	+0.013	0.025	-0.007				+0.013	0.030	-0.007	+0.018	0.043	-0.013
6	10	+0.010	0.019	-0.006	+0.015	0.036	-0.010				+0.015	0.041	-0.010	+0.023	0.051	-0.015
10	14	+0.010	0.020	-0.008	+0.018	0.041	-0.010				+0.018	0.046	-0.010	+0.025	0.058	-0.018
14	18	+0.010	0.023	-0.008	+0.018	0.041	-0.010				+0.018	0.046	-0.010	+0.025	0.064	-0.018
18	24	+0.013	0.028	-0.010	+0.020	0.048	-0.013				+0.020	0.053	-0.013	+0.030	0.076	-0.020
24	30	+0.013	0.030	-0.010	+0.020	0.048	-0.013	+0.020	0.053	-0.013	+0.020	0.058	-0.013	+0.030	0.084	-0.020
30	40	+0.015	0.033	-0.010	+0.025	0.061	-0.015	+0.025	0.066	-0.015	+0.025	0.079	-0.015	+0.041	0.102	-0.025
40	50	+0.015	0.036	-0.010	+0.025	0.061	-0.015	+0.025	0.071	-0.015	+0.025	0.086	-0.015	+0.041	0.127	-0.025
50	65	+0.018	0.046	-0.013	+0.030	0.069	-0.018	+0.030	0.082	-0.018	+0.030	0.107	-0.018	+0.046	0.157	-0.030
65	80	+0.018	0.048	-0.013	+0.030	0.074	-0.018	+0.030	0.094	-0.018	+0.030	0.119	-0.018	+0.046	0.183	-0.030
80	100	+0.023	0.061	-0.015	+0.035	0.094	-0.023	+0.035	0.112	-0.023	+0.036	0.150	-0.023	+0.056	0.213	-0.036

Force and shrink fits

APPENDIX G

DRILLED HOLE TOLERANCE (UNDER NORMAL SHOP CONDITIONS)

STANDARD DRILL SIZE				TOLERANCE IN DECIMALS	
DRILL SIZE				PLUS	MINUS
Number	Fraction	Decimal	Metric (MM)		
80		0.0135	0.3412	0.0023	
79		0.0145	0.3788	0.0024	
—	1/64	0.0156	0.3969	0.0025	
78		0.0160	0.4064	0.0025	
77		0.0180	0.4572	0.0026	
76		0.0200	0.5080	0.0027	
75		0.0210	0.5334	0.0027	
74		0.0225	0.5631	0.0028	
73		0.0240	0.6096	0.0028	
72		0.0250	0.6350	0.0029	
71		0.0260	0.6604	0.0029	
70		0.0280	0.7112	0.0030	.0005
69		0.0292	0.7483	0.0030	
68		0.0310	0.7874	0.0031	
—	1/32	0.0312	0.7937	0.0031	
67		0.0320	0.8128	0.0031	
66		0.0330	0.8382	0.0032	
65		0.0350	0.8890	0.0032	
64		0.0360	0.9144	0.0033	
63		0.0370	0.9398	0.0033	
62		0.0380	0.9652	0.0033	
61		0.0390	0.9906	0.0033	
60		0.0400	1.0160	0.0034	
59		0.0410	1.0414	0.0034	
58		0.0420	1.0668	0.0034	
57		0.0430	1.0922	0.0035	
56		0.0465	1.1684	0.0035	
—	3/64	0.0469	1.1906	0.0036	
55		0.0520	1.3208	0.0037	
54		0.0550	1.3970	0.0038	
53		0.0595	1.5122	0.0039	
—	1/16	0.0625	1.5875	0.0039	
52		0.0635	1.6002	0.0039	
51		0.0670	1.7018	0.0040	
50		0.0700	1.7780	0.0041	
49		0.0730	1.8542	0.0041	.001
48		0.0760	1.9304	0.0042	
—	5/64	0.0781	1.9844	0.0042	
47		0.0785	2.0001	0.0042	
46		0.0810	2.0574	0.0043	
45		0.0820	2.0828	0.0043	
44		0.0860	2.1844	0.0044	
43		0.0890	2.2606	0.0044	
42		0.0935	2.3622	0.0045	
—	3/32	0.0937	2.3812	0.0045	
41		0.0960	2.4384	0.0045	
40		0.0980	2.4892	0.0046	
39		0.0995	2.5377	0.0046	
38		0.1015	2.5908	0.0046	
37		0.1040	2.6416	0.0047	
36		0.1065	2.6924	0.0047	
—	7/64	0.1094	2.7781	0.0047	

DRILLED HOLE TOLERANCE (UNDER NORMAL SHOP CONDITIONS)

No./Letter	Fraction	Decimal	Metric (MM)	PLUS	MINUS
35		0.1100	2.7490	0.0047	
34		0.1110	2.8194	0.0048	
33		0.1130	2.8702	0.0048	
32		0.1160	2.9464	0.0048	
31		0.1200	3.0480	0.0049	
—	1/8	0.1250	3.1750	0.0050	
30		0.1285	3.2766	0.0050	
29		0.1360	3.4544	0.0051	
28		0.1405	3.5560	0.0052	
—	9/64	0.1406	3.5719	0.0052	
27		0.1440	3.6576	0.0052	
26		0.1470	3.7338	0.0052	
25		0.1495	3.7886	0.0053	
24		0.1520	3.8608	0.0053	
23		0.1540	3.9116	0.0053	
—	5/32	0.1562	3.9687	0.0053	
22		0.1570	3.9878	0.0053	
21		0.1590	4.0386	0.0054	
20		0.1610	4.0894	0.0054	
19		0.1660	4.2164	0.0055	
18		0.1695	4.3180	0.0055	
—	11/64	0.1719	4.3656	0.0055	
17		0.1730	4.3942	0.0055	
16		0.1770	4.4958	0.0056	.001
15		0.1800	4.5720	0.0056	
14		0.1820	4.6228	0.0057	
13		0.1850	4.6990	0.0057	
—	3/16	0.1875	4.7625	0.0057	
12		0.1890	4.8006	0.0057	
11		0.1910	4.8514	0.0057	
10		0.1935	4.9276	0.0058	
9		0.1960	4.9784	0.0058	
8		0.1990	5.0800	0.0058	
7		0.2010	5.1054	0.0058	
—	13/64	0.2031	5.1594	0.0058	
6		0.2040	5.1816	0.0058	
5		0.2055	5.2070	0.0059	
4		0.2090	5.3086	0.0059	
3		0.2130	5.4102	0.0059	
—	7/32	0.2187	5.5562	0.0060	
2		0.2210	5.6134	0.0060	
1		0.2280	5.7912	0.0061	
A		0.2340	5.9436	0.0061	
—	15/64	0.2344	5.9531	0.0061	
B		0.2380	6.0452	0.0061	
C		0.2420	6.1468	0.0062	
D		0.2460	6.2484	0.0062	
E	1/4	0.2500	6.3500	0.0063	
F		0.2570	6.5278	0.0063	
G		0.2610	6.6294	0.0063	
—	17/64	0.2656	6.7469	0.0064	
H		0.2660	6.7564	0.0064	
I		0.2720	6.9088	0.0064	.002
J		0.2770	7.0358	0.0065	
K		0.2810	7.1374	0.0065	
—	9/32	0.2812	7.1437	0.0065	
L		0.2900	7.3660	0.0066	
M		0.2950	7.4930	0.0066	
—	19/64	0.2969	7.5406	0.0066	

DRILLED HOLE TOLERANCE (UNDER NORMAL SHOP CONDITIONS)

STANDARD DRILL SIZE				TOLERANCE IN DECIMALS	
DRILL SIZE				PLUS	MINUS
Letter	Fraction	Decimal	Metric (MM)		
N		0.3020	7.6708	0.0067	
—	5/16	0.3125	7.9375	0.0067	
O		0.3160	8.0264	0.0068	
P		0.3230	8.2042	0.0068	
—	21/64	0.3281	8.3344	0.0068	
Q		0.3320	8.4328	0.0069	
R		0.3390	8.6106	0.0069	
—	11/32	0.3437	8.7312	0.0070	
S		0.3480	8.8392	0.0070	
T		0.3580	9.0932	0.0071	
—	23/64	0.3594	9.1281	0.0071	
U		0.3680	9.3472	0.0071	
—	3/8	0.3750	9.5250	0.0072	
V		0.3770	9.5758	0.0072	
W		0.3860	9.8044	0.0072	
—	25/64	0.3906	9.9219	0.0073	
X		0.3970	10.0838	0.0073	
Y		0.4040	10.2616	0.0073	
—	13/32	0.4062	10.3187	0.0074	
Z		0.4130	10.4902	0.0074	.002
	27/64	0.4219	10.7156	0.0075	
	7/16	0.4375	10.1125	0.0075	
	29/64	0.4531	11.5094	0.0076	
	15/32	0.4687	11.9062	0.0077	
	31/64	0.4844	12.3031	0.0078	
	1/2	0.5000	12.7000	0.0079	
	33/64	0.5156	13.0968	0.0080	
	17/32	0.5312	13.4937	0.0081	
	35/64	0.5469	13.8906	0.0081	
	9/16	0.5625	14.2875	0.0082	
	37/64	0.5781	14.6844	0.0083	
	19/32	0.5927	15.0812	0.0084	
	39/64	0.6094	15.4781	0.0084	
	5/8	0.6250	15.8750	0.0085	
	41/64	0.6406	16.2719	0.0086	
	21/32	0.6562	16.6687	0.0086	
	43/64	0.6719	17.0656	0.0087	
	11/16	0.6875	17.4625	0.0088	
	45/64	0.7031	17.8594	0.0088	
	23/32	0.7187	18.2562	0.0089	
	47/64	0.7344	18.6532	0.0090	
	3/4	0.7500	19.0500	0.0090	
	49/64	0.7656	19.4469	0.0091	
	25/32	0.7812	19.8433	0.0092	
	51/64	0.7969	20.2402	0.0092	
	13/16	0.8125	20.6375	0.0093	
	53/64	0.8281	21.0344	0.0093	
	27/32	0.8437	21.4312	0.0094	
	55/64	0.8594	21.8281	0.0095	
	7/8	0.8750	22.2250	0.0095	.003
	57/64	0.8906	22.6219	0.0096	
	29/32	0.9062	23.0187	0.0096	
	59/64	0.9219	23.4156	0.0097	
	15/16	0.9375	23.8125	0.0097	
	61/64	0.9531	24.2094	0.0098	
	31/32	0.9687	24.6062	0.0098	
	63/64	0.9844	25.0031	0.0099	
	1	1.0000	25.4000	0.0100	

INDEX

INDEX

Aligned dimensioning, 150
Allowance. *See* Dimensioning
Alloys, 188
Alphabet of lines, 22
Angularity, 181
Application, job, 330
Arrowheads, 151, 153
Auxiliary views
    front auxiliary, 122
    projecting, 122
    side auxiliary, 122
    top auxiliary, 122

Basic hole system. *See* Dimensioning
Basic size. *See* Dimensioning
Bend allowances, 145, 146, 147, 335, 336, 337, 338
Bilateral tolerancing. *See* Dimensioning
Bisecting
    a line, 26
    an angle, 27
Bolt circle. *See* Dimensioning
Borders, 20
Bosses, 200
Brittleness of material, 188
Brush, drafting, 9

Calipers, 238, 239, 244
Cams
    design and layout, 257, 262, 263
    dimensioning, 264, 265
    displacement diagram, 258, 259, 260
    followers, 256, 257
    timing, 264
Careers in drafting, 316, 317, 318, 319, 320, 321
Casting, 194, 195
Chamfer. *See* Dimensioning

Change procedure. *See* Engineering Dept.
Checking process. *See* Engineering Dept.
Class of fit. *See* Fasteners
Cleaning pad, 10
Clearance. *See* Dimensioning
Compass, 11
Concentricity, 181
Conductivity of material, 188
Cope. *See* Castings
Core. *See* Castings
Corrosion resistance of material, 188
Cotter pin. *See* Fasteners
Counterbore holes, 157
Countersunk holes, 157
Crest of threads. *See* Fasteners
Curves, irregular, 7
Cutting plane line, 92

Descriptive geometry
    angles between two surfaces, 118
    edge view, 116
    fold lines, 110, 111, 112, 126, 127, 128
    notations, 110, 111, 112
    plane surface, 115
    point view, 114
    true length, 113
    true shape, 117
    visibility of lines, 119
Developments
    base line, 126
    fold lines, 110, 111, 112, 126, 127, 128
    layout, 129, 130, 131, 133
    radial line, 135
    tabs, 132, 134
    triangulation, 141, 142, 143
    true length, 139, 140, 141, 142

Dimensioning
    allowance, 166
    baseline dimensioning, 173, 174
    basic hole system, 167
    basic size, 167
    bilateral tolerancing, 168
    bolt circle, 160
    chamfer, 155
    change procedure, 154
    clearance, 167
    diameter, 153
    fits, kinds of, 168
    holes, 154, 157, 170
    knurling, 155
    leaders, 151
    limits, 164
    lines, 151
    location dimensions, 156
    maximum material condition (MMC), 165, 166
    radius, 153
    reference, 155
    rules of, 151
    size, 156
    tolerance, 164, 168
    unilateral tolerance, 168
Divider, 6, 238
Drag. *See* Castings
Drafting machines, 2
Drawings, kinds of
    assembly, 268, 269
    design layout, 268
    detail, 268, 270, 272, 273, 274, 275, 276, 277
    purchased parts, 268
    subassembly, 268, 271
Drilled holes, 157, 345, 346, 347
Drilling operation, 192
Ductivity of material, 188

Elasticity of material, 188
Ellipse, 40, 41
Employer-Employee agreement. *See* Engineering Dept.
Engineering Dept.
    attitude, 331
    change procedure, 310
    checking process, 308
    employer-employee agreement, 314
    invention agreement, 331
    numbering systems, 311
    organization, 308, 309
    parts list, 311, 312, 313
    practices, 308
    project flow chart, 314, 315
Eraser, 10
Extrusion, 190

Fasteners
    callouts, 219, 220
    class of fits, 215, 220
    crest, 214

    double thread, 216
    major diameter, 214
    minor diameter, 214
    pitch of, 214
    representation, 217
    root, 214
    tap and die, 216
    thread form, 215
    thread gauge, 221
    thread terms, 214
Fatigue limit, 188
Feature control symbols, 178, 179, 180, 181, 182, 186
Ferrous metal, 189
Fillets, 66, 102, 199
Finish surfaces, 158, 159, 160
Fits, kinds of. *See* Dimensioning
Flask. *See* Castings

Grinding operation, 193
Guidelines, 18

Heat treatment, 189
Holder, lead, 8
Hexagon, 30

Instrument sets, 6
Intersections, 63, 64, 68
Interview, job, 327, 328, 329
Invention agreement. *See* Engineering Dept.
Isometric views
    axis, 72
    box construction, 74
    centering, 80
    circles, 81, 83
    curves, 78
    cylinders, 85
    non-isometric lines, 73
    radii, 82
    skeleton construction, 76

Justifying columns, 285

Keys, 168, 230
Key ways, 230
Knurling. *See* Dimensioning

Lead, 8
Lead holder, 8
Lettering
    guidelines, 18
    mechanical, 280, 281, 282, 283, 286, 287, 288, 289
    spacing, 19
    strokes, 18
    style, 19
Limits. *See* Dimensioning
Lock washers. *See* Fasteners
Lugs, 200, 201

Machine screw. *See* Fasteners
Machine tool operations, 192

Major diameter. *See* Fasteners
Malleability of material, 188
Maximum material condition. *See* Dimensioning
Metals, 188
Metallurgy, 188
Microinch. *See* Finished surface
Micrometer, 240, 241, 242, 243
Milling operation, 194
Millimeter, 237
Minor diameter. *See* Fasteners
Modifiers, 182
Molding board. *See* Castings
Multi-views
    centering, 62
    numbering, 64
    rules for, 60, 61

Nominal size. *See* Dimensioning
Non-ferrous metal, 189
Numbering system. *See* Engineering Dept.

Ogee curves, 42, 43, 44, 45
Order of strokes. *See* Lettering

Pad, 200
Paddle, 11
Paper, size, 13, 20
Parallel straightedge, 3
Parallism, 180
Parts list. *See* Engineering Dept.
Pencils, 8
Perpendicularity, 180
Perspective drawing
    one point, 292, 298, 299
    terms, 297
    three point, 295, 296
    two point, 294, 300, 301, 302, 303, 304, 305, 306
Pitch-thread. *See* Fasteners
Planing, 193
Plasticity of material, 183
Protractor, 7

Reamed hole, 157
Resumé writing, 323, 324, 325, 326
Ribs, 200
Riser. *See* Castings
Rounds, 66, 102, 199
Root-thread. *See* Fasteners
Runout, 67, 102

Scale
    architectural, 14
    civil engineers, 15
    decimal, 15
    mechanical, 14
    metric, 15, 16
Seams, 148
Section lining. *See* Section views

Section views
    assembly, 101
    breaks, 103
    full, 94
    half, 95
    non-section parts, 102
    offset, 97
    practices, 103, 104, 107, 232
    removed, 100
    revolved, 98
    ribs/webs, 105
    rotated (revolved), 98
    section lining, 93
    thinwall, 102
Shading, 88, 89, 90
Shield-erasing, 9
Shrink rule, 201
Size-paper, 13, 20
Spotface, 157
Spring-bow compass, 11
Springs
    compression, 249, 250
    extension, 251
    how to draw, 250, 252, 253, 254
    kinds of, 247
    winding direction, 248
Sprue. *See* Castings
Strength of material, 188
Symmetry, 182

Tangent arcs, 46, 47, 48, 49, 50, 51, 52, 53
Tangent points, 38, 39, 67, 68
Tap and die. *See* Fasteners
Template, 6
Thread form. *See* Fasteners
Thread gauge. *See* Fasteners
Thread root. *See* Fasteners
Tolerance. *See* Dimensioning
Total indicator reading (TIR), 182
Toughness of material, 188
Triangle, 4, 5
True positioning, 183, 184, 185, 186
T-square, 3
Turning operation, 193

Undercut, 234
Unidirectional dimensioning, 150
Unilateral tolerancing. *See* Dimensioning

Vernier, 244, 245

Web, 200
Weight of material, 192
Welding-fusion
    joints, 204, 205
    symbols, 204, 206, 207, 208
Where to look for a job, 322
Whiteprinter, 116